General Science

BOOK THREE

Page 72 write

work of roots
Root growth
osmosis
Soil erosion

Page 73 Draw.

work of roots
osmosis
Soil erosion
Preventing.

General Science

BOOK THREE

by Charles Windridge

edited by Patrick J. Kenway, B.Sc.

illustrated by Barry Davies

SCHOFIELD & SIMS LTD · HUDDERSFIELD

0 7217 3553 3

0 7217 3558 4 Net edition

First edition 1959
Reprinted 1959, 1960, 1961, 1962, 1963, 1964, 1965, 1966, 1968

Second (revised) edition 1969
Reprinted 1970, 1971, 1973, 1974

Third (revised) edition 1977

Reprinted 1977, 1978

GENERAL SCIENCE is a series of five books.
Book One [0 7217 3551 7]
Book Two [0 7217 3552 5]
Book Three [0 7217 3553 3]
Book Four [0 7217 3554 1]
Book Five [0 7217 3555 X]

Printed in England by
Henry Garnett & Co Ltd, Rotherham and London

Author's Note

REVISED EDITION

This book is a completely up-dated and fully revised edition of General Science, Book Three, which was first published in 1959.

The complete rèvision of this book and the other books in the series ensures that they are fresh in their approach and that the series as a whole will satisfy the needs of pupils in this day and age—the Space Age. The revision has been carried out with a proper regard for the latest developments in scientific research and thought, the recent trends in the methodology of science teaching, and the ever-changing pattern of technological advancement, and in the awareness that science is not just a body of knowledge but also a set of methods and a way of thinking. Much of the revision has been concerned with those modifications that have been made necessary by the omission of obsolescent material, the introduction of SI metric units, new discoveries, changes in examination requirements, etc.; but the series still retains those familiar characteristics which, as very many reprints over the years clearly indicate, have made it very popular and established it as being thoroughly reliable and adequate.

This series provides a course in general and combined science that is complete, but in no way overloaded, for pupils of average ability within the 11 to 16 age-range. It is essentially a study of natural science—physics, chemistry, biology, astronomy, geology, meteorology, etc.—and its applications in technology and other fields. The fifth book in this series caters for those pupils who are taking preliminary technical courses or examinations in general science of the standard of the Certificate of Secondary Education, and it provides a sound groundwork for the important minority who will take trade courses and advanced courses in technology later on.

The course is presented as an integrated and solid core of indispensable and basic facts and techniques combined with coherent topics that will promote purposeful activities, encourage individual research and satisfy modern examination requirements. There are opportunities for pupils to acquire and use knowledge and skills in both real and realistic situations. They are made to understand that the ability to handle scientific situations, rather than the mere mental storage of scientific facts, is the proper end-product of scientific education.

An important feature of the books in this series is that both the diagrams and the text are almost entirely self-explanatory, and this, coupled with the tremendous variety and the wide coverage in the contents, means that most pupils could work from them without very much assistance from the teacher. Therefore, in addition to their usefulness as textbooks of the traditional type, the books are very suitable for use with mixed ability classes and for homework, projects and individual study.

The practical work is straightforward and homely, so that, quite often, it can be performed with simple or improvised apparatus. The diagrams are examples on which pupils can base their own drawings. The exercises, given at the end of the books, are intended to supplement the pupils' practical work and records of experiments by providing extra practice in written work, drawing, the use of scientific terms, calculations, etc., and, of course, they are useful for revision purposes. The units of measurement are SI (Le Système International d' Unités) and decimal notation is used almost exclusively throughout.

Full consideration has been given to the various reports and recommendations of such bodies as the Certificate of Secondary Education and University Examination Boards, the Association for Science Education, the Council of Technical Examining Bodies, the Royal Society, the Royal Institute of Chemistry, etc.

C.W.

Metric Units

THE STANDARD INTERNATIONAL SYSTEM OF UNITS

Le Système International d' Unités, which is known here as the Standard International System of Units, or, more simply, as S I, is the official measuring system of the United Kingdom, and all the units of measurement that are used in this series of books are part of this system except in a few instances where, for historical or some other special reasons, it is necessary to do otherwise.

Some of the S I fundamental, derived and supplementary units are:

physical quantity	*unit*	*symbol*
length	metre	m
mass	kilogram	kg
time	second	s
temperature	kelvin	K
temperature	degree Celsius (customary unit)	°C
electric current	ampere	A
luminous intensity	candela	cd
area	square metre	m^2
volume	cubic metre	m^3
velocity	metre per second	m/s
acceleration	metre per second per second	m/s^2
force	newton	N
heat, work and energy	joule	J
power	watt	W

The above-mentioned units and some others not already mentioned are defined and fully explained at appropriate places in the text.

As far as is possible all fractions are expressed in decimal notation so that full benefit is derived from a system of units whose multiples and sub-multiples can be so easily expressed in that same notation.

Further information about S I units and conventions can be obtained from the publications that are issued from time to time by such bodies as the Royal Society, the Association for Science Education, the British Standards Institution, the Royal Institute of Chemistry, Her Majesty's Stationery Office, the Council of Technical Examining Bodies, the Department of Education and Science, etc. The booklets *The Use of S I Units in the Early and Middle Years of Schooling* and *S I Units, Signs, Symbols and Abbreviations for Use in School Science,* both published by the Association for Science Education, are particularly helpful publications.

Contents

1 Magnets

Magnetic attraction

Bring a small *compass* close to various objects made of iron or steel. Nails, pins, screws, needles, penknife blades and screwdrivers are suitable objects. The compass needle, which is a *magnet,* turns and points towards the objects.

Magnets attract objects made of iron or steel. You are bound to have seen the toy *horseshoe* magnets that are used for picking up small iron or steel objects, such as pins and needles.

Magnetic and non-magnetic materials

Hold various small objects, such as a needle, a pin, a piece of copper wire, a brass screw, a wooden block, a glass tube, etc., close to a magnet. The magnet attracts only those objects which are made of iron or steel.

Iron, steel, *cobalt, nickel* and certain alloys are attracted by a magnet; they are called *magnetic materials. Ferrites,* which are mixtures that include *iron oxide,* used in the aerials and coils of transistor radios, are also magnetic materials. However, most materials are not magnetic.

Separating screws

Bring a magnet into contact with a quantity of brass and iron screws. Only the iron screws stick to the magnet.

Magnetism acts through materials

Place thin sheets of paper, glass, rubber and copper, in turn, between a *bar magnet* and an iron nail. You notice that the magnetism acts through these materials. Use the magnet to pick up an iron nail lying at the bottom of a dish of water.

The poles of a magnet

Dip a bar magnet into iron filings. The iron filings cluster together at the ends of the magnet and so show the positions of the two *poles*.

Produce the same effect by using a *lodestone.* Lodestone is a form of iron oxide which has magnetic properties.

The properties of magnets

Suspend a bar magnet from its centre with a length of cotton or silk. The magnet turns and comes to rest with its ends pointing in a direction which is roughly north-south. Check this with a compass needle.

Bring together the ends of two suspended magnets. Make a table to show the attraction and repulsion between the poles.

Poles Used	*Result*
A north pole brought near a south pole	
A north pole brought near a north pole	
A south pole brought near a north pole	
A south pole brought near a south pole	

North and south poles

A magnet has a *north-seeking pole* at one end and a *south-seeking pole* at the other. For short, a north-seeking pole is called a *north pole,* and a south-seeking pole is called a *south pole.* When freely suspended, the north pole of a magnet points to the Earth's *Magnetic North Pole.* The north pole of a magnet repels the north pole and attracts the south pole of another magnet. This law is often stated briefly as *"like poles repel and unlike poles attract".*

The shapes of magnets

Magnets are made in various shapes. Horseshoe magnets retain their magnetism longer than bar magnets. The reason for this is given on page 10. The magnetism of *ball-ended magnets* is "concentrated" in the balls. Powerful *alnico* magnets retain their magnetism for long periods. These are made from special alloys (see page 62) of aluminium, cobalt, nickel, copper and iron.

Magnetic fields

Place a bar magnet and a horseshoe magnet beneath a large sheet of paper or glass. Sprinkle iron filings on the paper. Gently tap the paper, and the iron filings will take up positions on the *lines of force* of the magnets. Notice the shapes of the *magnetic fields.* The lines of force join opposite poles.

The field of magnetic force

There is an invisible *field of magnetic force* around a magnet. The strength of this field is greater in the regions near the poles of the magnet than in regions some distance from the poles. It is for this reason that magnets pick up iron or steel objects only when they are brought close to them.

MAGNETIC ATTRACTION

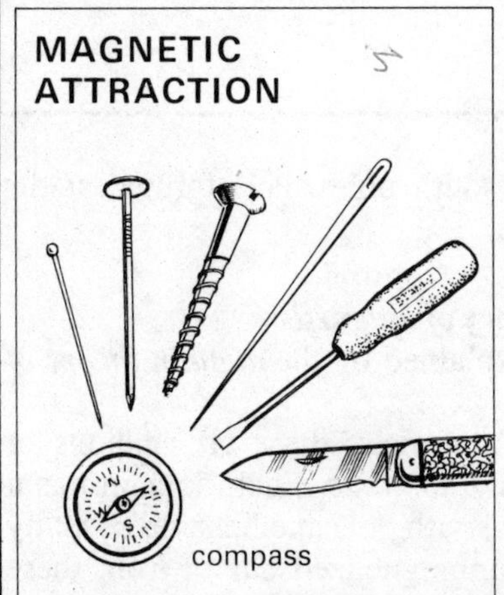

MAGNETIC AND NON-MAGNETIC MATERIALS

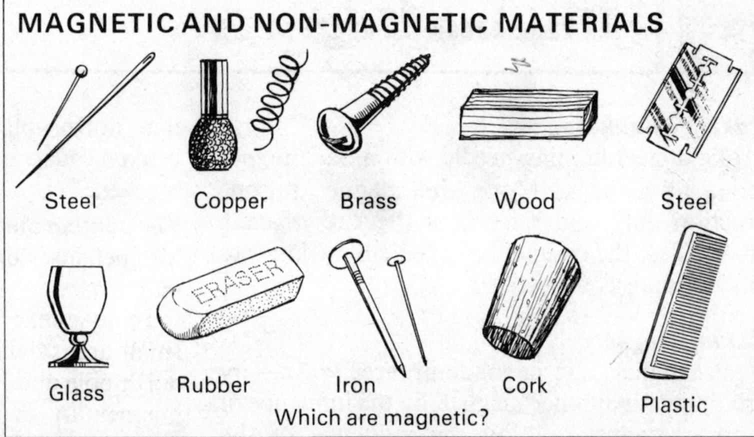

Which are magnetic?

SEPARATING SCREWS

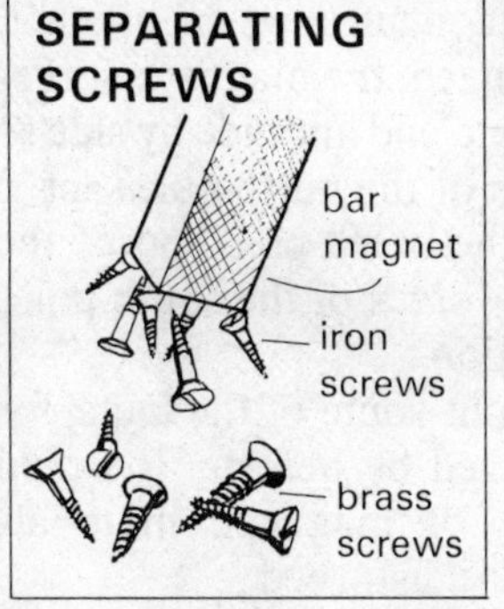

MAGNETS

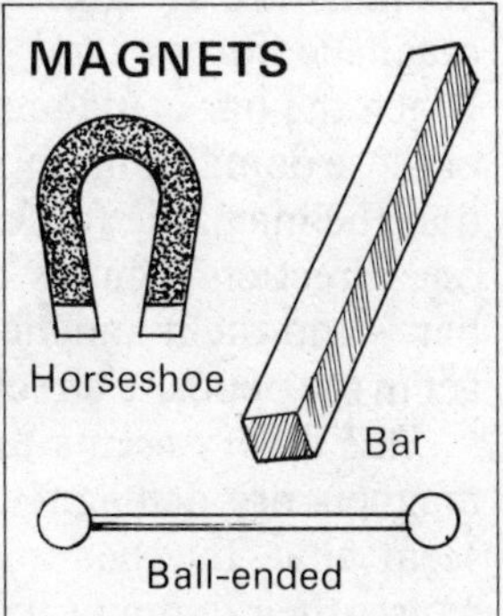

MAGNETISM ACTS THROUGH MATERIALS

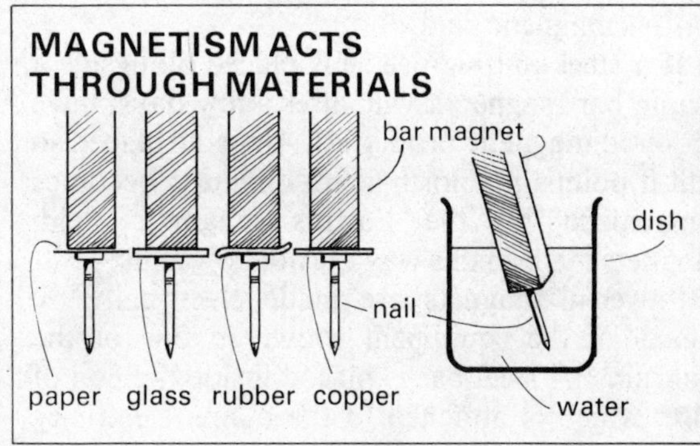

PROPERTIES OF MAGNETS

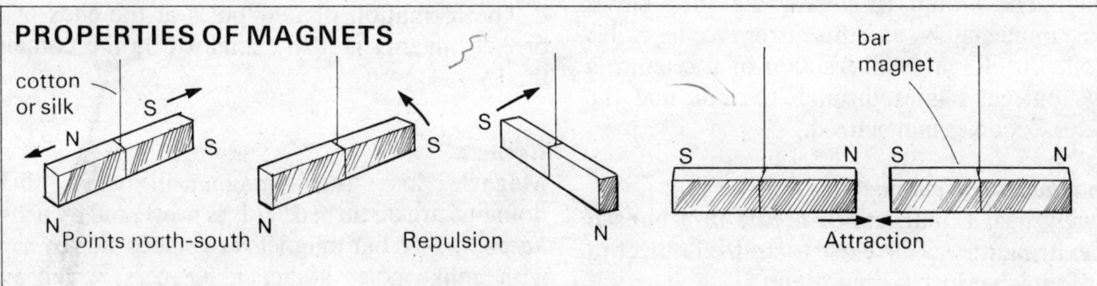

Like poles repel and unlike poles attract

THE POLES OF A MAGNET

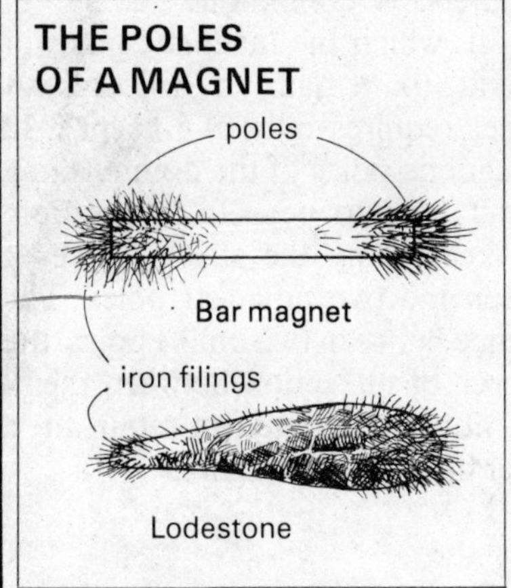

MAGNETIC FIELDS

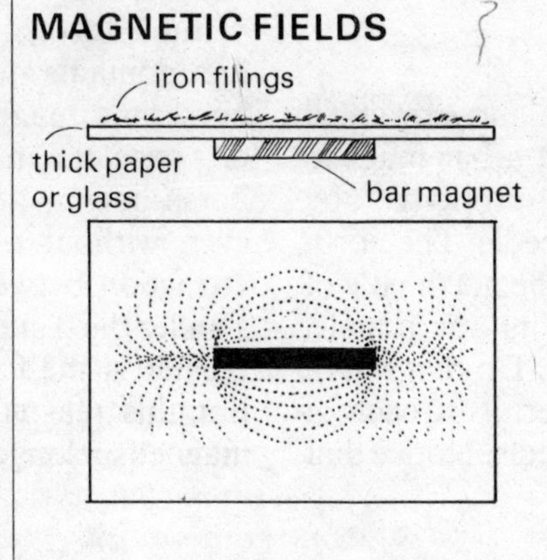

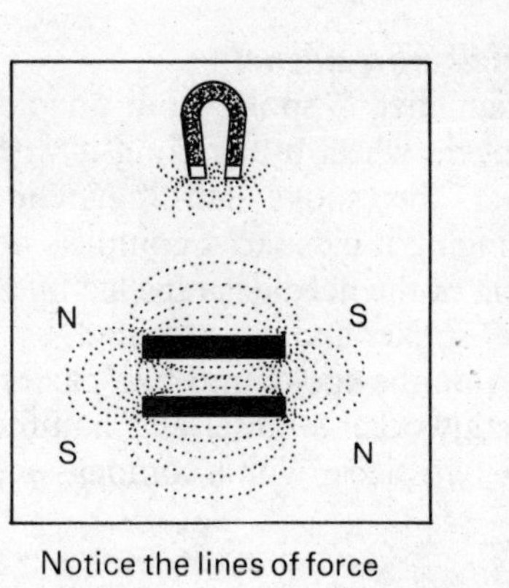

Notice the lines of force

2 Making Magnets

Making a weak magnet

Stroke a steel knitting-needle with a bar magnet about fifty times. Move the magnet in one direction only and raise it at the end of each stroke. Test the magnetized needle by using it to pick up small iron nails.

Making magnets

Weak magnets can be made in three ways—by stroking with another magnet, by the influence of another magnet and by the influence of the Earth's magnetic field.

If a steel knitting-needle is placed alongside a strong bar magnet, it will, after a few days, have acquired magnetic properties. A needle placed so that it points in a north-south direction becomes magnetized by the Earth's magnetic field. Magnets made in this way are not powerful.

Powerful magnets are made electrically by means of the equipment shown in one of the diagrams. A steel bar is placed inside the coil of wire, which is attached to the mains electricity supply. The switch is closed, the fuse blows almost immediately, and the current to the coil is cut off. But, for a small fraction of a second, a heavy current passes through the coil and the steel bar becomes magnetized.

Demagnetizing a magnet

Strongly heat a magnetized needle in a bunsen flame. Bring the needle close to an iron object to show that it has lost its magnetism.

Magnets are demagnetized when they are strongly heated or hammered.

Breaking a magnet

Magnetize a spoke from an old and unwanted bicycle wheel by stroking it with a bar magnet. Test the spoke for magnetic properties by bringing it close to a compass needle. The north pole of the needle is repelled by the north pole of the spoke and is attracted by its south pole. Divide the spoke into two pieces. Do this with a metalworker's chisel and hammer. Test each of the two pieces with a compass needle. Notice that new north and south poles have formed at the broken ends.

The domain theory of magnetism

Magnetism is explained by the *domain theory of magnetism.*

In magnetic materials, there are millions of small areas called *domains*. Each domain has a north pole and a south pole and behaves as a tiny magnet. In an unmagnetized bar of iron, these domains are arranged in such a way that their magnetic forces tend to neutralize each other. When the bar is magnetized, the magnetic forces of these domains lie end to end and side by side so that the magnetic forces of the north poles act in one direction—this is the north-pole end of the bar—and the magnetic forces of the south poles act in the opposite direction.

This theory seems to fit some of the facts, for magnets are demagnetized by heating. It would seem that the heating of magnetic materials causes their domains to be disturbed.

The formation of new poles at the ends of a broken magnet is also explained by the domain theory.

Keepers

Magnets lose their magnetism when their domains are disturbed. This is prevented by using *keepers*. Two bar magnets are placed side by side with unlike poles adjacent. Keepers, which are small bars or rods of iron, are then placed at the ends of the magnets. A continuous "circle" of metal is formed, in which the magnetic poles of the domains tend to remain in position. A horseshoe magnet requires only one keeper. In any case, the magnetic poles of the domains in a horseshoe magnet tend to be held in position, even without a keeper, by the strong force of attraction between the two adjacent poles. The smaller the distance between two unlike poles, the greater is the force of attraction between them. For this reason horseshoe magnets retain their magnetism longer than do bar magnets.

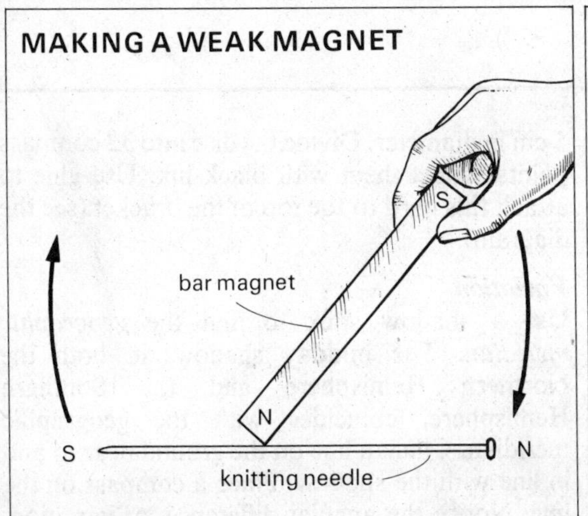
MAKING A WEAK MAGNET
S
bar magnet
N
S
N
knitting needle

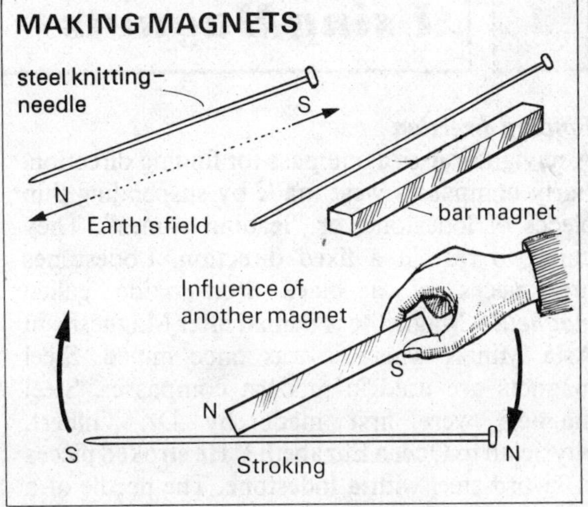
MAKING MAGNETS
steel knitting-
needle
S
N
Earth's field
bar magnet
Influence of
another magnet
S
N
S
N
Stroking

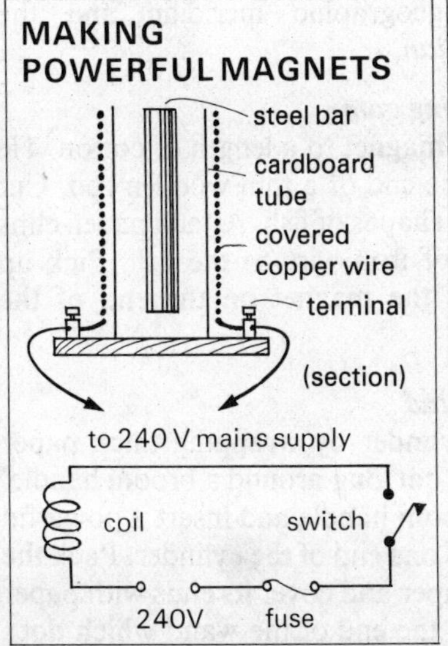
MAKING
POWERFUL MAGNETS
steel bar
cardboard
tube
covered
copper wire
terminal
(section)
to 240 V mains supply
coil
switch
240V
fuse

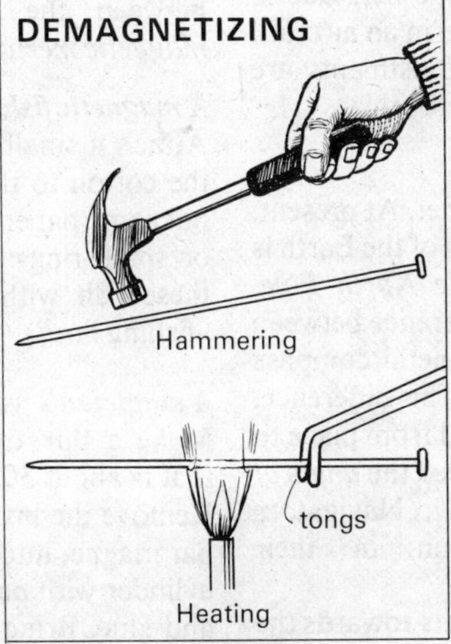
DEMAGNETIZING
Hammering
tongs
Heating

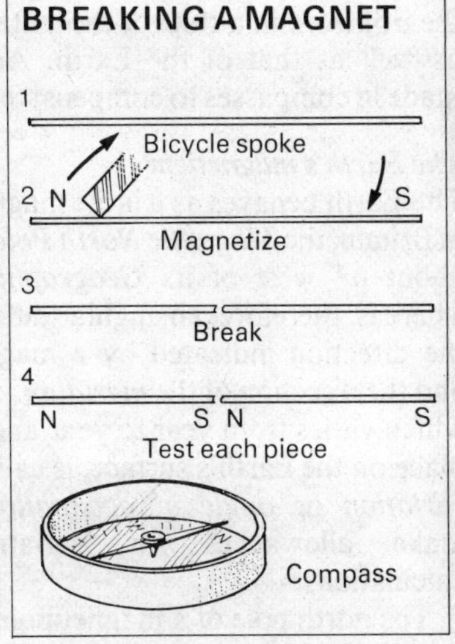
BREAKING A MAGNET
1
Bicycle spoke
2 N
S
Magnetize
3
Break
4
N
S N
S
Test each piece
Compass

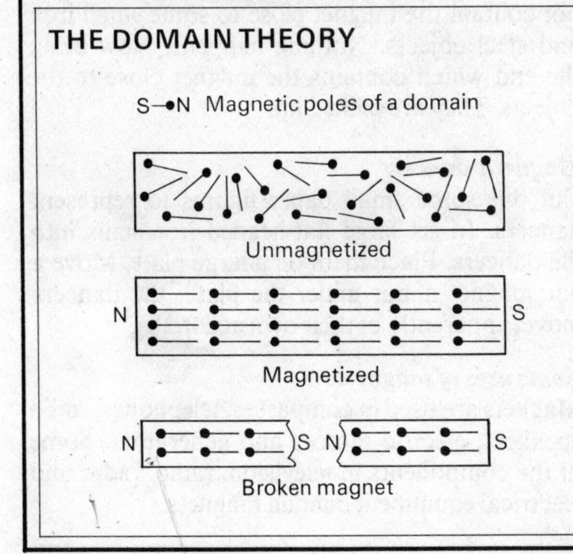
THE DOMAIN THEORY
S—•N Magnetic poles of a domain
Unmagnetized
N
S
Magnetized
N
S N
S
Broken magnet

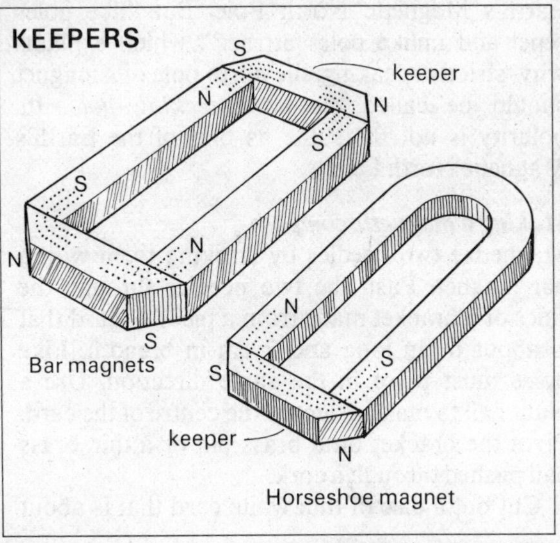
KEEPERS
S
keeper
N
N
S
S
N
N
N
S
Bar magnets
S
keeper
N
Horseshoe magnet

3 Using Magnets

Finding direction
A navigator uses a compass for finding direction. Early compasses were made by suspending thin pieces of lodestone, or "leading stones". They came to rest in a fixed direction. Lodestones are pieces of a black iron oxide called *magnetite.* Magnetite is named after Magnesia, in Asia Minor, where it was once mined. Steel magnets are used in modern compasses. Steel magnets were first made by Dr. Gilbert, physician to Queen Elizabeth I. He stroked pieces of sword steel with a lodestone. The needle of a compass is affected by the magnetic field due to the ironwork in a ship, or the metal in an aircraft, as well as that of the Earth. Adjustments are made in compasses to compensate for this.

The Earth's magnetism
The Earth behaves as a large magnet. At present, in Britain, the *Magnetic North Pole* of the Earth is about 6° west of its *Geographic North Pole.* There is, therefore, an angular difference between the direction indicated by a magnetic compass and the direction of the *meridian*. This difference, which varies from year to year and from place to place on the Earth's surface, is called the *angle of variation* or *angle of declination.* Navigators make allowances for variation in their calculations.

The north pole of a magnet points towards the Earth's Magnetic North Pole. But "like poles repel and unlike poles attract", which explains why, strictly speaking, the north pole of a magnet should be called the *north-seeking pole.* Its polarity is not the same as that of the Earth's Magnetic North Pole.

Making a magnetic compass
Magnetize two needles by stroking them with a bar magnet. Push the two needles through the sides of a bracket made from a piece of card that is about 6 cm long and 2 cm in breadth. Like poles must point in the same direction. Use a blunt nail to make a dent in the centre of the card. Pivot the bracket on a brass pin or a thin brass nail pushed through a cork.

Cut out a disc of thin white card that is about 5 cm in diameter. Divide the disc into 32 compass points. Label them with black ink. Use glue to attach this card to the top of the bracket (see the diagram).

Variation
Use a shadow stick to find the *geographic meridian.* The midday shadow, in both the Northern Hemisphere and the Southern Hemisphere, coincides with the geographic meridian. Chalk a line on the ground near to and in line with the shadow. Place a compass on the line. Notice the angular difference, or variation, between the geographic meridian and the *magnetic meridian.*

A magnetic fishing game
Attach a small magnet to a length of cotton. Tie the cotton to the end of a thin wooden rod. Cut out some paper shapes of fish. Attach paper-clips or small rings of iron wire to the fish. Pick up these fish with the magnet on the end of the "fishing line".

A magician's wand
Make a thin cylinder by wrapping thick paper that is about 30 cm long around a broom handle. Remove the broom handle and insert a powerful bar magnet into one end of the cylinder. Pack the cylinder with paper and cover its ends with paper and glue. Bring the end of the wand which does not contain the magnet close to some small iron and steel objects. Nothing happens. Now bring the end which contains the magnet close to the objects. They are picked up.

Magnetic dancers
Cut out some small paper figures to represent dancers. Insert large flat-headed iron nails into the dancers. Place them on a large plate. Move a bar magnet about under the plate; the dancers move, apparently of their own accord.

Some uses of magnets
Magnets are used in compasses, telephones, radio speakers, electric motors and generators. Some of the components in television, radio, radar and electrical equipment contain magnets.

FINDING DIRECTION

MAGNETIC NORTH

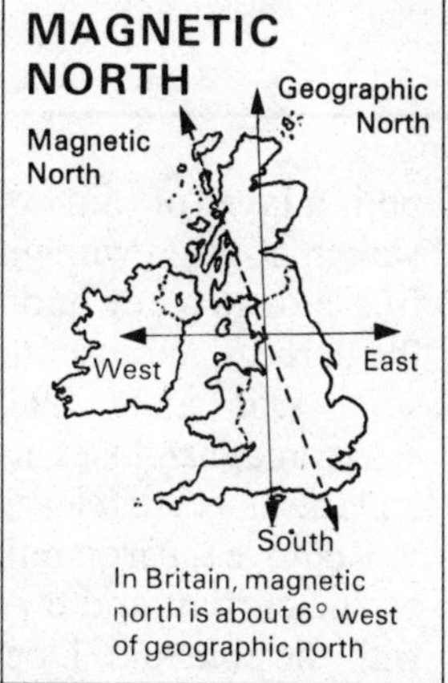

VARIATION

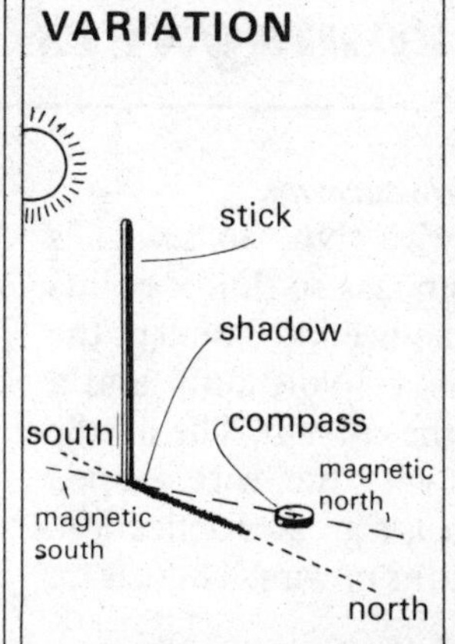

MAGNETIC FISHING

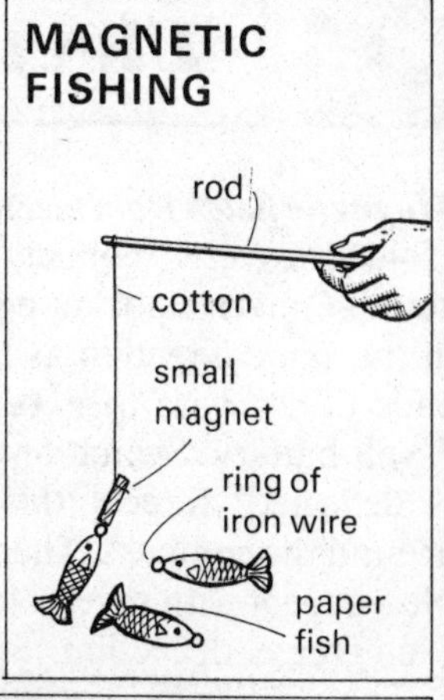

MAGNETIC COMPASS

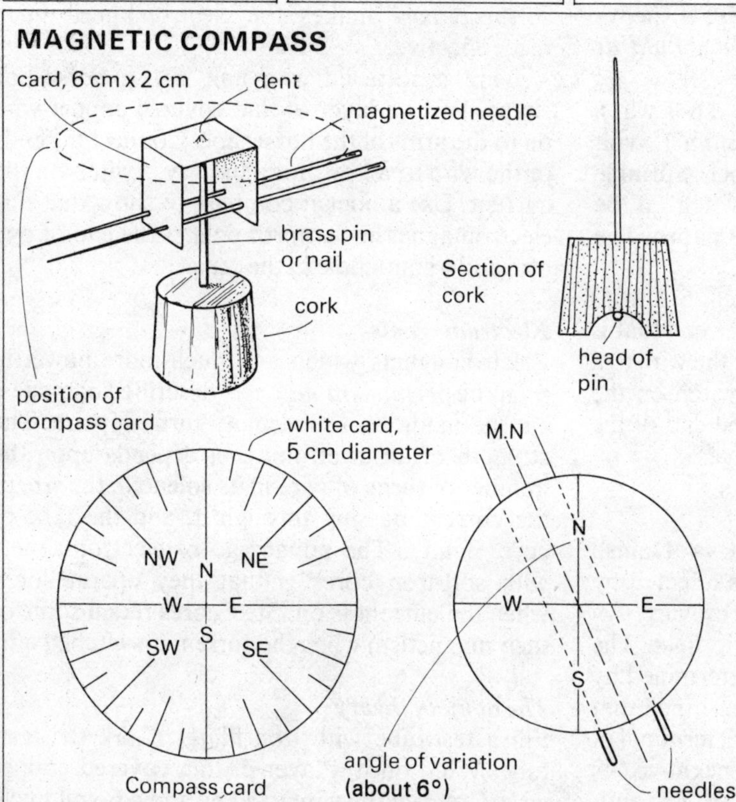

COMPASS CARD WITH 32 POINTS

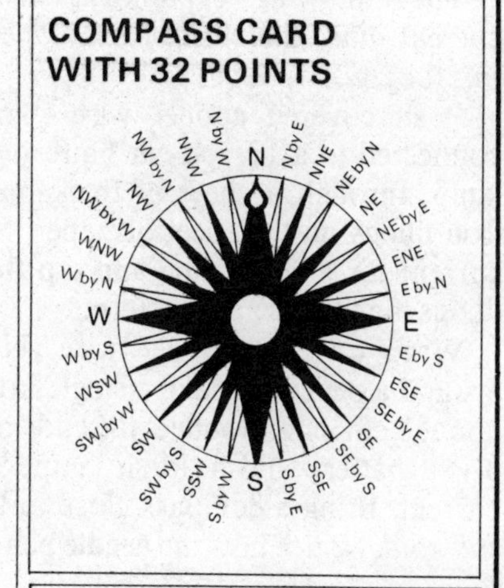

MAGNETIC DANCERS

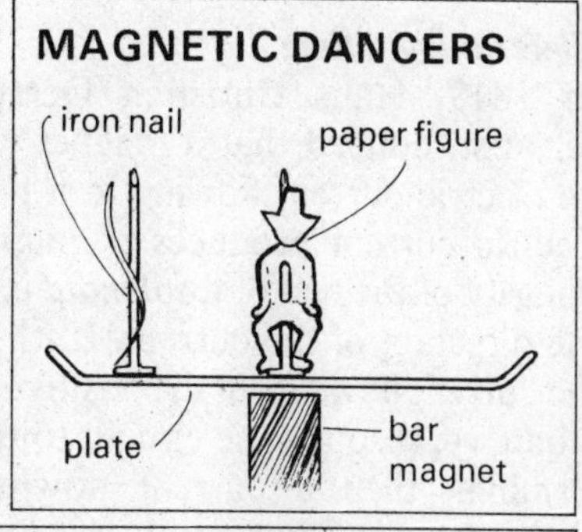

SOME USES OF MAGNETS

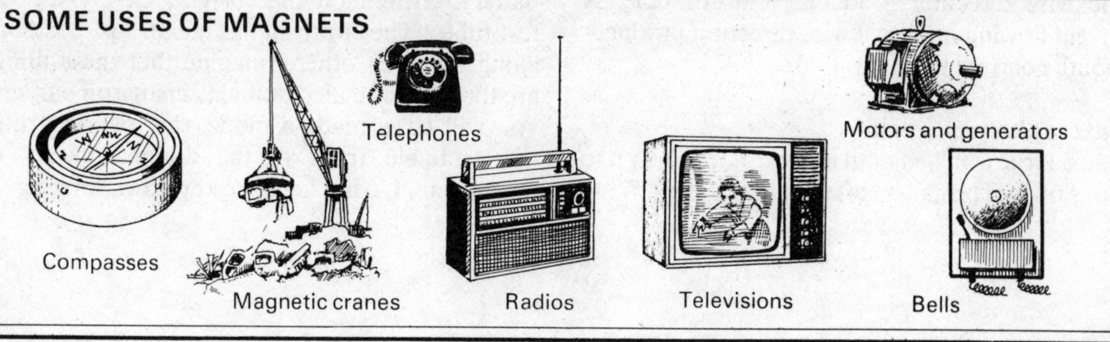

4 Electromagnets

Magnetic fields from electric currents

Hold a covered copper wire (No. 26 s.w.g. is suitable) over a pocket compass so that it points in the same direction as the needle. Connect the ends of the wire to a *key,* a 3-volt lamp and a 3-volt battery. Switch on the *current.* The needle is deflected. Repeat this with the wire looped around the compass. There is a greater deflection. Now repeat with several turns of wire. There is an even greater deflection.

The following experiment needs a heavy current, and, therefore, *your teacher should do this for you.*

Push covered copper wire (No. 26 s.w.g.), connected to a 12-volt car battery and a 12-volt lamp, through a sheet of thick paper. Sprinkle iron filings on the sheet of paper. Switch on the current for a short time and tap the paper. The filings make a circular pattern.

Wrap covered copper wire (No. 26 s.w.g.) around a pencil to make a single-layer coil that is about 8 cm long. Connect the ends of the wire to a 3-volt battery and a 3-volt lamp. Switch on the current. Bring a compass close to the end of the *solenoid.* Notice how the needle behaves.

Electromagnetism

In 1819, Hans Christian Oersted, a Danish scientist, noticed that a magnet was affected by an electric current flowing in a nearby wire. An electric current produces a magnetic field. The polarity of the end of a *solenoid* is determined by the direction of the current. If it is assumed that the current flows from the positive (+) terminal of a battery, through the circuit, to the negative (−) terminal, then a current flowing in an anti-clockwise direction produces a north pole. A current flowing in a clockwise direction produces a south pole (see the diagrams).

Making electromagnets

Heat a large iron nail until it is red-hot. Allow it to cool slowly. The nail is softened in this way. Wind on it a layer of thin covered copper wire. Fix the wire in position with gummed paper. Connect the free ends to a key and a 3-volt battery. Switch on the current, and use the *electromagnet* to pick up small iron nails. Switch off the current. The electromagnet loses its magnetism immediately and the nails are released.

Remove the iron nail from the solenoid. Switch on the current and try to pick up small iron nails with the solenoid. Does it pick up the nails? The soft-iron *core* makes the electromagnet much more effective.

Bend a softened iron nail into a horseshoe shape. Wind a layer of thin covered copper wire on to the arms of the horseshoe. Connect the ends of the wire to a key and a battery. Switch on the current. Use a pocket compass to show that this electromagnet has a north pole at the end of one arm and a south pole at the other.

Electromagnets

Electromagnets which are much more powerful than the *permanent magnets* described on page 8 can be made by using many turns of wire. The strength of an electromagnet depends upon the number of turns of wire in its solenoid, the size of the current passing through it, and the type of core, if any. The advantage of electromagnets with soft-iron cores is that they operate only when the current is on. Steel cores retain some of their magnetism when the current is switched off.

The domain theory

Fill a test-tube with iron filings. Cork the test-tube. Wind on it a layer of thin covered copper wire. Connect the wire to a key and a 6-volt cycle battery. Switch on the current. Gently tap the test-tube. The iron filings take up positions alongside each other. Imagine that these filings are the regions called domains in an iron bar, and you will have made a model that demonstrates the probable truth of the domain theory of magnetism. Do not keep the key down for long.

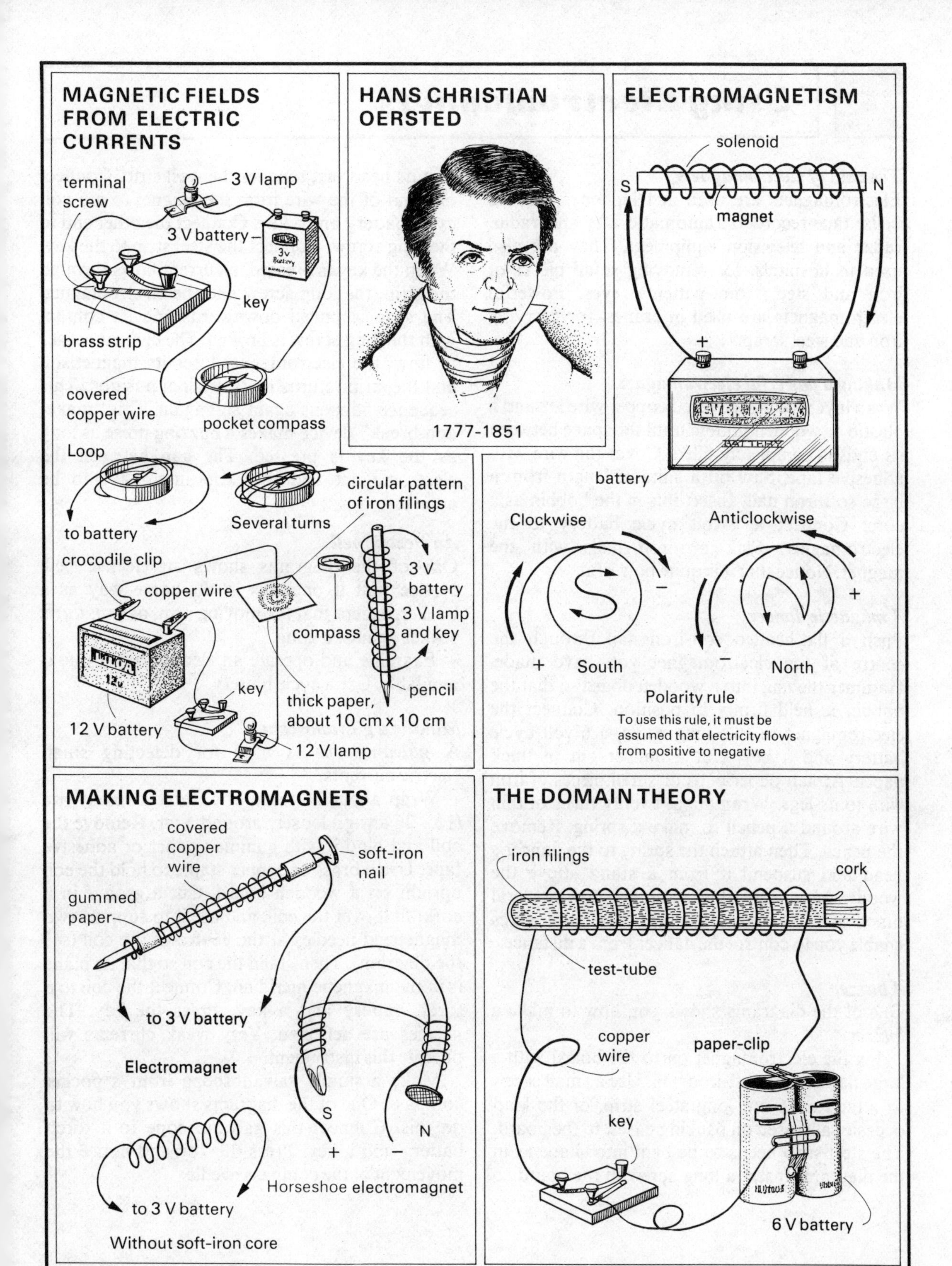

MAGNETIC FIELDS FROM ELECTRIC CURRENTS
terminal screw
3 V lamp
3 V battery
3v Battery
key
brass strip
covered copper wire
pocket compass
Loop
to battery
Several turns
circular pattern of iron filings
crocodile clip
copper wire
3 V battery
3 V lamp and key
compass
key
12v
12 V battery
thick paper, about 10 cm x 10 cm
pencil
12 V lamp
HANS CHRISTIAN OERSTED
1777-1851
ELECTROMAGNETISM
solenoid
S
N
magnet
+
−
EVER READY
BATTERY
battery
Clockwise
Anticlockwise
S
N
South
North
Polarity rule
To use this rule, it must be assumed that electricity flows from positive to negative
MAKING ELECTROMAGNETS
covered copper wire
soft-iron nail
gummed paper
to 3 V battery
Electromagnet
S
N
+
−
Horseshoe electromagnet
to 3 V battery
Without soft-iron core
THE DOMAIN THEORY
iron filings
cork
test-tube
copper wire
paper-clip
key
6 V battery

5 Using Electromagnets

The uses of electromagnets

Electromagnets are used in telephones, electric bells, tape-recorders, automatic lifts and radio, radar and television equipment. They are also used in hospitals, for removing small pieces of iron and steel from patients' eyes. Powerful electromagnets are used in cranes, for carrying iron and steel scrap.

Making a powerful electromagnet

Wrap layers of thin covered copper wire around a plastic or wooden bobbin until the space between its ends is completely filled. Cover the wire with adhesive tape. Saw off a suitable length from a large soft-iron nail. Insert this in the bobbin as a core. Connect a 6-volt cycle battery to the electromagnet. Pick up iron nails with the magnet. Notice that it is quite powerful.

A magnetic dancer

Push a flat-headed soft-iron nail through the centre of the electromagnet you have made. Hammer the nail into a wooden board so that the bobbin is held firmly in position. Connect the electromagnet to two long wires, a 6-volt cycle battery and a key. Cut a dancer out of thick paper. Attach paper-clips or small pieces of iron wire to its legs. Wrap about twenty turns of thin wire around a pencil to make a spring. Remove the pencil. Then attach the spring to the dancer's head and suspend it from a stand above the wooden base. Press the key up and down several times and the figure dances. The long wires enable you to control the dancer from a distance.

A buzzer

One of the diagrams shows you how to make a *buzzer*.

Fix the electromagnet on to the board with a large flat-headed soft-iron nail. Use a small screw to attach a 15 cm long steel strip, of the kind occasionally used on packing cases, to the board. The steel strip needs to be bent into shape as in the diagram. Insert a long screw in the board so that its head just touches the steel strip. Connect one end of the wire from the magnet to a 6-volt cycle battery and a key. Connect the other end to the long screw. Connect the steel strip to the key. When the key is pressed, a current flows through the strip, the long screw and the electromagnet. The strip is pulled downwards and its contact with the long screw is broken. The current ceases to flow, the electromagnet loses its magnetism, and the strip returns to its original position. This sequence happens again and again. This "make-and-break" device makes a buzzing noise as long as the key is pressed. The gap between the electromagnet and the strip may have to be adjusted.

An electric bell

One of the diagrams shows an electric bell. Notice that it operates in the same way as a buzzer, except that the moving arm, or *armature*, strikes against a gong.

Examine and operate an electric bell if one is available. Use a cycle battery.

Making a galvanoscope

A *galvanoscope* is used for detecting small electric currents.

Wrap about 20 turns of covered copper wire (No. 26 s.w.g.) loosely around a jar. Remove the coil and bind it with gummed paper or adhesive tape. Use a brass or copper staple to hold the coil upright on a wooden board. Cut a groove in a cork, fit it over the coil, and use it to support two magnetized needles at the centre of the coil (see the diagram). Then stand the coil so that its plane is in the magnetic meridian. Connect the coil to a torch battery and a key. Press the key. The needles are deflected. Very weak currents will operate this instrument.

Make a simple galvanoscope from a pocket compass. One of the diagrams shows you how to do this. Connect this galvanoscope to a torch battery and a key. Press the key and notice the movement of the compass needle.

MAKING A POWERFUL ELECTROMAGNET

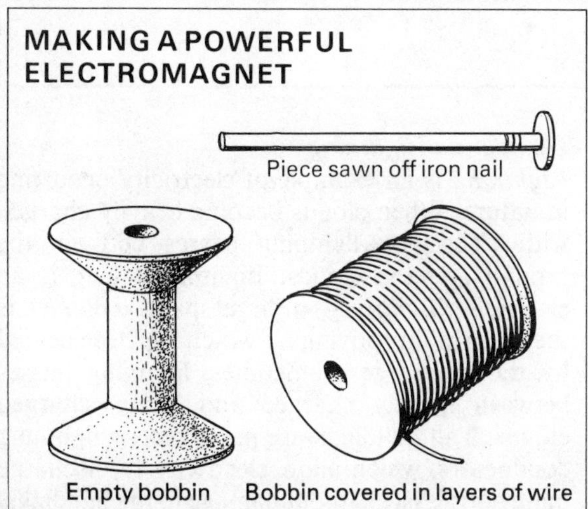

MAGNETIC DANCER

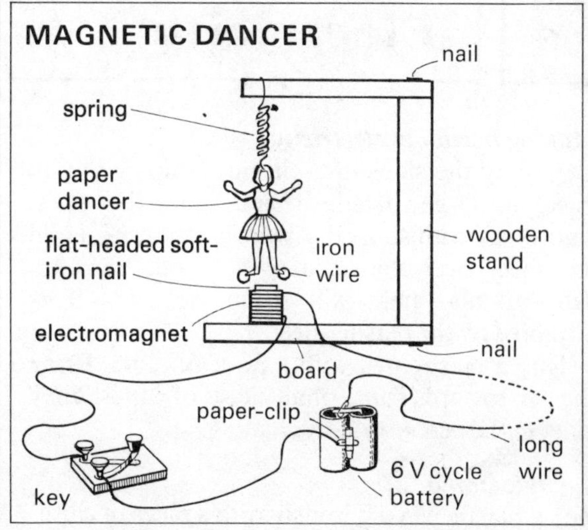

BUZZER

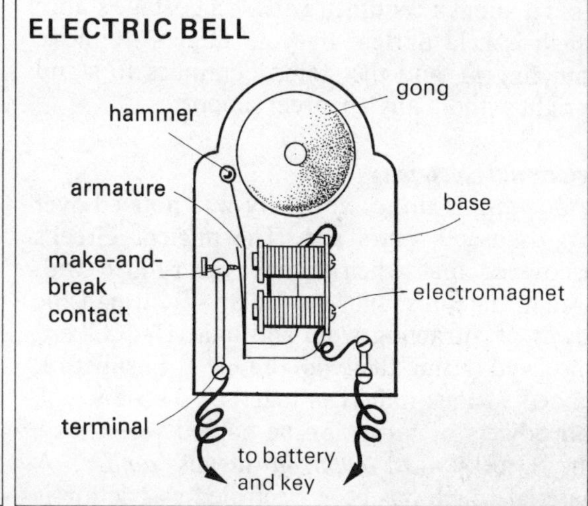

ELECTRIC BELL

MAKING A GALVANOSCOPE

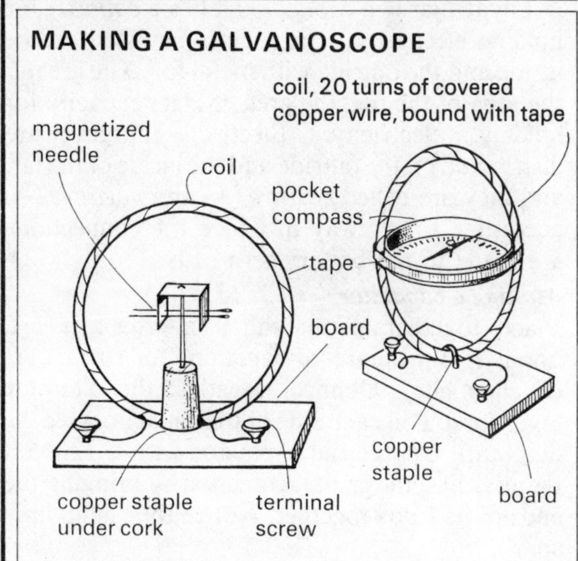

HOSPITAL ELECTROMAGNET

Removing a steel splinter from a patient's eye

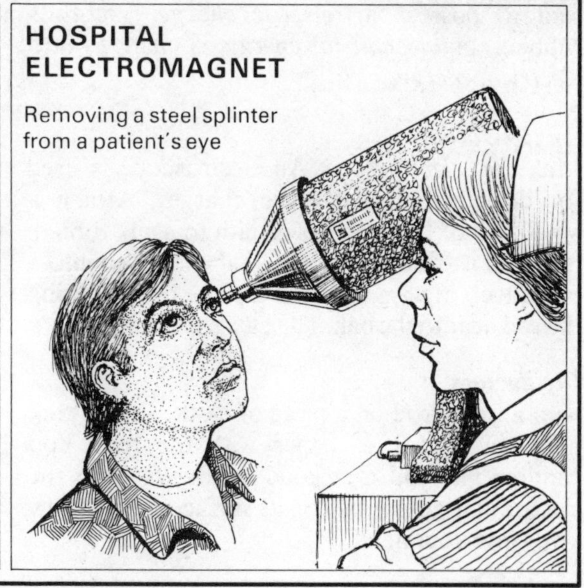

6 Electricity

Making frictional electricity
Cut away the sides of a plastic shopping bag to make a large plastic sheet. Rub the sheet vigorously with some dry woollen material. Hold the sheet over the head of a friend. Some of her—or his—hair will stand on end. It is attracted by the plastic sheet.

Rub a plastic pen with a piece of cloth. Bring the pen towards some small pieces of paper. They cling to the pen.

A magic stamp
Rub a plastic pen vigorously with a piece of cloth. Hold it about a centimetre above a postage stamp which is held upright by your fingers. Remove your fingers, and the stamp continues to stand upright without any apparent support.

Frictional electricity
Frictional, or *static,* electricity was noticed over two thousand years ago. The ancient Greeks discovered that when *amber* was rubbed with certain materials, such as fur or silk, it had the power of attracting wool and hair. Dr. Gilbert, who lived during the reign of Queen Elizabeth I, rubbed various materials together. Those which had powers of attraction he named "electrics". The Greek word *elektron* means *amber.* A material which has been electrified by friction is said to possess an *electric charge.* Vigorous rubbing, or *mechanical energy,* is changed into *electrical energy.*

An electroscope
Make an *electroscope.* An electroscope is used for detecting small electric charges. Attach a small ball of tissue paper or pith to a silk thread. Suspend it from a bent glass tube inserted into a cork. Rub a glass rod with a piece of silk. Bring the rod near to the ball. The ball is attracted.

Conductors
Rub a metal rod on a piece of cloth. Try to pick up small pieces of paper with the rod. You cannot. The rod is a good *conductor,* and the electric charges formed on its surface rapidly flow away into your hand.

Thunder and lightning
Lightning is an example of electricity occurring in nature. When clouds become heavily charged with electricity, lightning passes between the ground and the clouds. Lightning, then, is an electrical *discharge*—a large spark. *Thunder* is the noise of the moving air which has been heated by the discharge. Sometimes, lightning passes between heavily charged and lightly charged clouds. Tall buildings are protected by lightning conductors, which allow electricity to discharge quite harmlessly without damage to the buildings.

Benjamin Franklin, an American scientist, flew a kite during a thunderstorm. When the silk thread on the kite became wet, it acted as a conductor. He used the electricity carried by this conductor to charge a *Leyden jar* and concluded that lightning and electricity were one and the same thing. It was as well for Franklin that he did not hold the thread in his hand. He would have been killed if the electricity had discharged through his body.

A Leyden jar
Connect a Leyden jar to a 12-volt battery for a short time. Disconnect the battery and connect the *leads* from the jar to a galvanoscope. Its needle is deflected.

Capacitors and capacity
A Leyden jar is a *capacitor;* it has a *capacity* for holding electricity. It is a glass jar lined on the inside and the outside with metal foil. The greater the area of the foil, the greater is its capacity for holding electricity. Electric charges are distributed on the outside and the inside of the jar, and they are called *positive* (+) and *negative* (–) according to the way in which the connections are made to the battery terminals.

Making a capacitor
Place together about thirty alternate layers, about 15 cm square, of paper and foil (from bars of chocolate). Connect together the alternate layers of foil on each side of the capacitor (see the diagram). Charge the capacitor with a 120-volt battery. Discharge the capacitor by bringing the ends of its leads together. Notice the "lightning" spark.

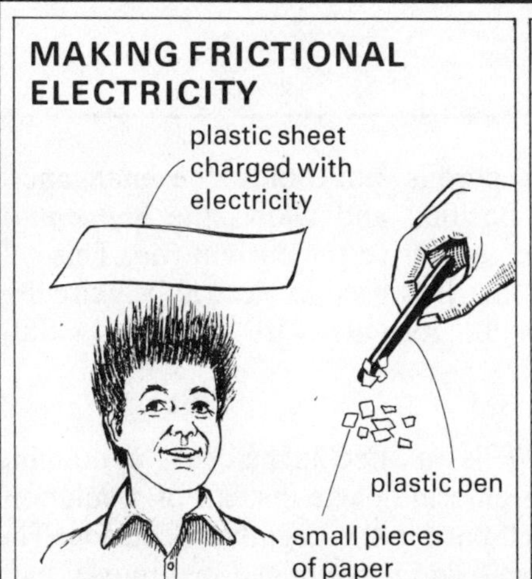
MAKING FRICTIONAL ELECTRICITY
plastic sheet charged with electricity
plastic pen
small pieces of paper

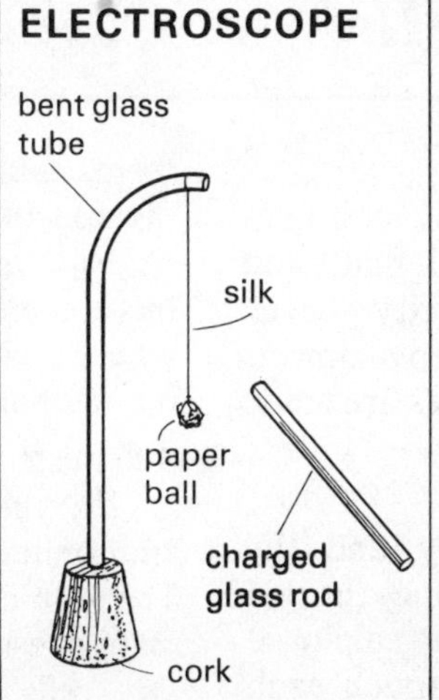
ELECTROSCOPE
bent glass tube
silk
paper ball
charged glass rod
cork

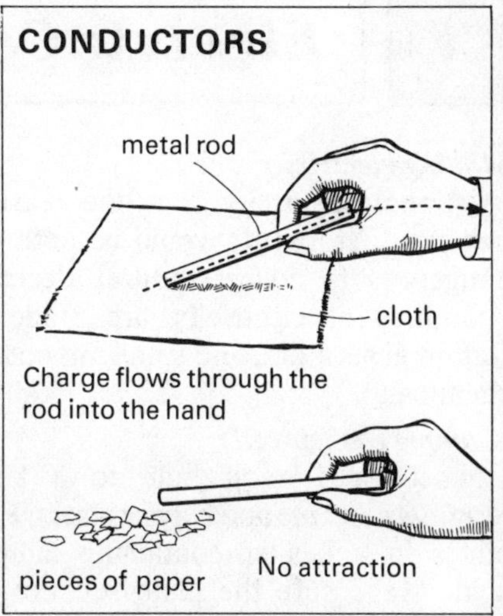
CONDUCTORS
metal rod
cloth
Charge flows through the rod into the hand
pieces of paper
No attraction

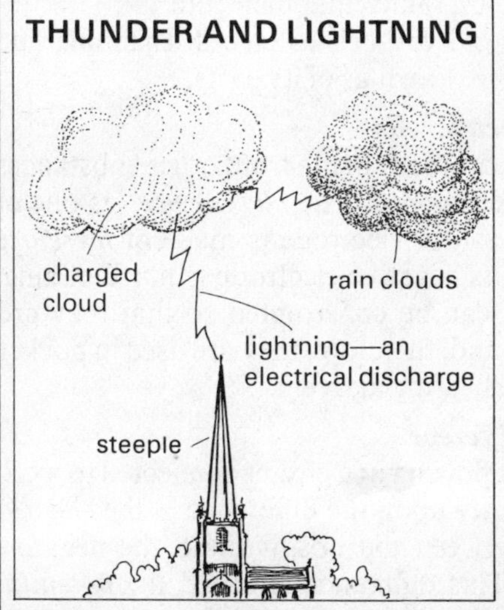
THUNDER AND LIGHTNING
charged cloud
rain clouds
lightning—an electrical discharge
steeple

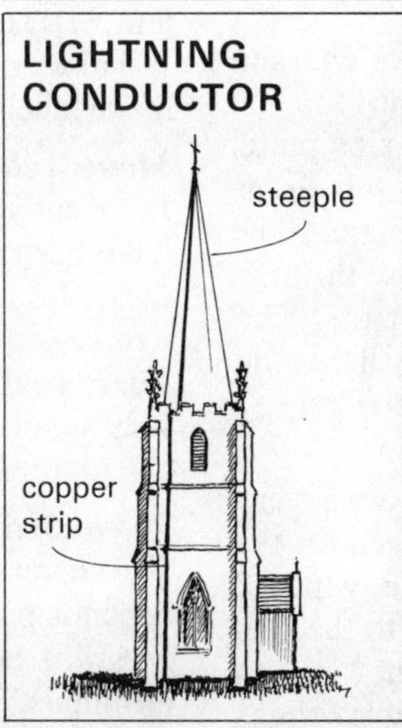
LIGHTNING CONDUCTOR
steeple
copper strip

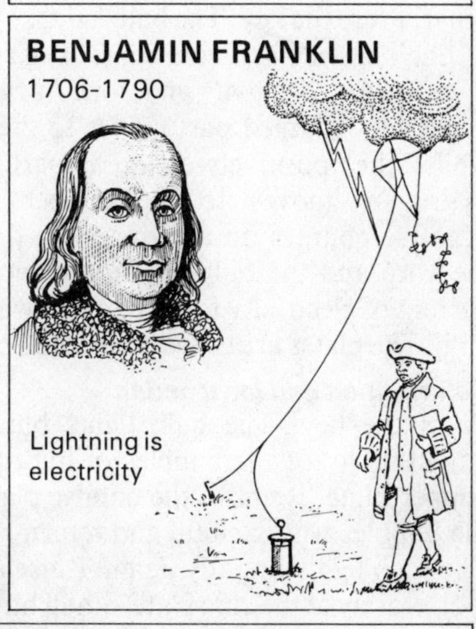
BENJAMIN FRANKLIN
1706-1790
Lightning is electricity

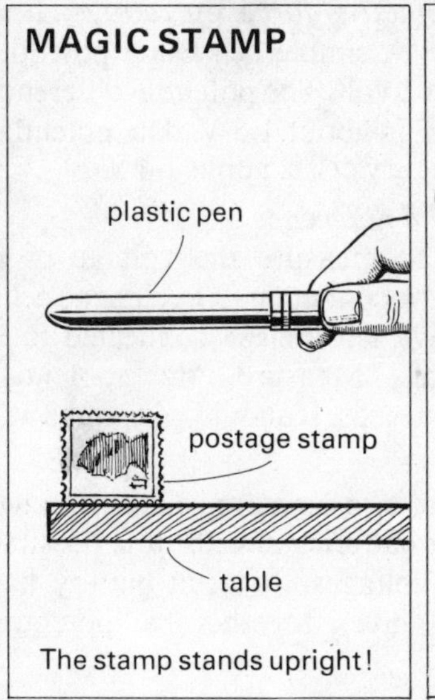
MAGIC STAMP
plastic pen
postage stamp
table
The stamp stands upright!

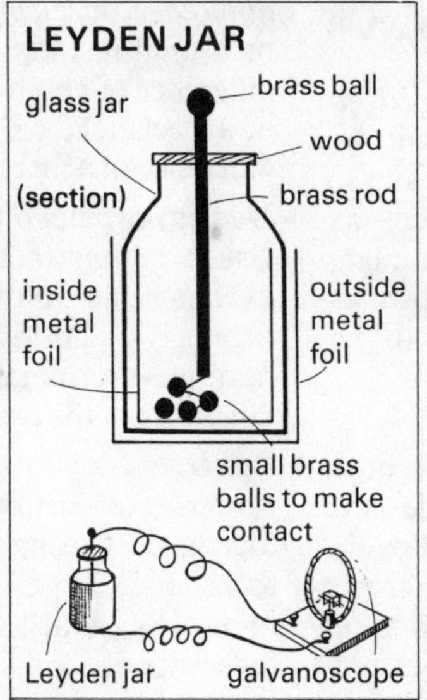
LEYDEN JAR
glass jar
brass ball
wood
(section)
brass rod
inside metal foil
outside metal foil
small brass balls to make contact
Leyden jar
galvanoscope

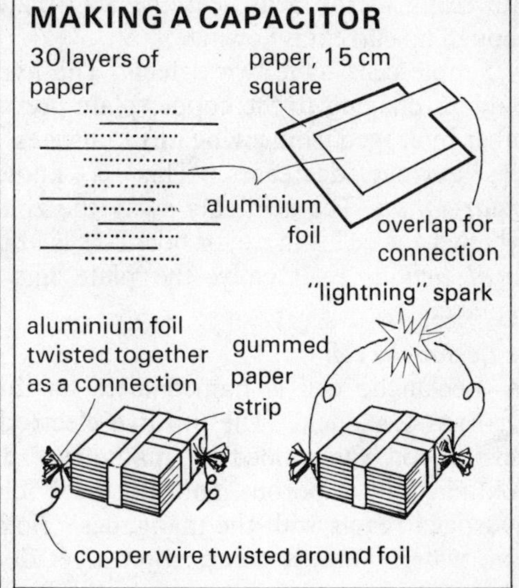
MAKING A CAPACITOR
30 layers of paper
paper, 15 cm square
aluminium foil
overlap for connection
"lightning" spark
aluminium foil twisted together as a connection
gummed paper strip
copper wire twisted around foil

7 Electric Cells

Making electricity

Frictional electricity is neither easily collected nor easily stored. It would be both difficult and dangerous to collect natural electricity. Large quantities of electricity are made by power-station generators, and small quantities are made chemically.

A simple electric cell

Connect a 1.5-volt bulb to a key and the terminals on *zinc* and *copper* plates. Place the two plates in a beaker containing dilute sulphuric acid. Make sure the plates do not touch each other. Press the key. The bulb lights.

Each molecule of acid consists of two charged parts, one positive and one negative. The negatively-charged part moves to the zinc plate, while the positively-charged part, which is hydrogen, moves to the copper plate. The negative charges on the zinc plate flow through the wire and the bulb to the copper plate. The current of electricity formed in this way lights the bulb. The plates are called *poles* or *electrodes*.

Polarization and local action

Press the key. The bulb lights but dims very quickly. Notice the bubbles of hydrogen on the copper plate. Remove the copper plate, wipe off the bubbles with a cloth, and return it to the cell. The bulb lights brightly again. Raise the key and look closely at the zinc plate. Small bubbles of gas are rising in the acid near the zinc plate even though no current is flowing.

Simple cells have two defects. The hydrogen bubbles clinging to the copper plate prevent the other hydrogen from giving up its charges, and so the current decreases. This is known as *polarization*. The acid eats away the zinc plate whether the cell is in use or not. This is known as *local action*. Eventually, the plate has to be replaced.

A Leclanché cell

A Leclanché cell is named after its inventor, Georges Leclanché. The positive electrode is a carbon rod surrounded by manganese dioxide contained in a porous pot. Positively-charged hydrogen reacts with the manganese dioxide to give water. Thus, a polarization layer does not form. The porous pot holds the manganese dioxide in position and allows the *ammonium chloride* free access to the carbon rod. This cell has a long life. It is used in electric bell circuits, where small currents are required only occasionally.

A dry cell

A *dry cell* is a Leclanché cell containing ammonium chloride paste instead of a solution. The zinc cell case is the negative electrode. The manganese dioxide is contained in a muslin bag. Dry cells are portable and are used in bicycle lamps, torches, radio equipment, etc.

Take an old dry cell to pieces. Examine and make labelled drawings of its parts.

Modern electric cells

In recent years, cells made from other substances have been developed. The *mercury cell* is one of these. Its positive electrode is made of *mercuric oxide,* and its negative electrode is made of zinc. These cells can be constructed so that they are very small, and, therefore, they are used in pocket calculators, hearing aids, etc.

Potential difference

The current flowing in a circuit connected to a cell depends partly upon the difference in the *electric pressure* between the positive and the negative terminals. This difference is called the *potential difference, electromotive force* or *voltage*. It is measured in *volts*. A simple cell has a potential difference of about 1 volt. The potential difference of a Leclanché cell is about 1.5 V. The potential difference of a mercury cell is about 1.4 V.

Measuring potential difference

Use a *voltmeter* to measure the voltage of a Leclanché cell. Now connect two Leclanché cells together. A positive terminal is connected to a negative terminal. Measure the potential difference of the two cells. It should be about 3 V.

Batteries

Batteries consist of two or more cells connected together. By using batteries of cells, it is possible to produce large voltages. A torch battery has two cells which give, together, a potential difference of 3 volts.

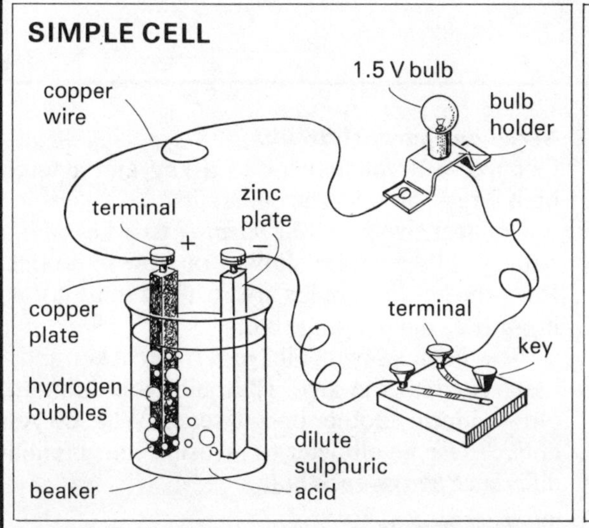
SIMPLE CELL
1.5 V bulb
copper wire
bulb holder
terminal
zinc plate
+
−
copper plate
terminal
key
hydrogen bubbles
dilute sulphuric acid
beaker

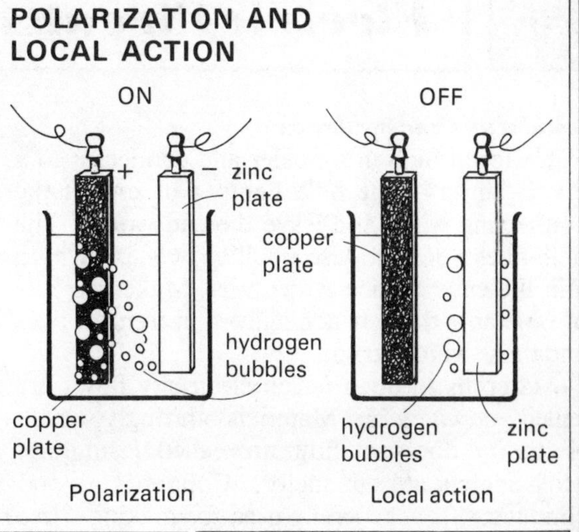
POLARIZATION AND LOCAL ACTION
ON
OFF
+
−
zinc plate
copper plate
hydrogen bubbles
copper plate
hydrogen bubbles
zinc plate
Polarization
Local action

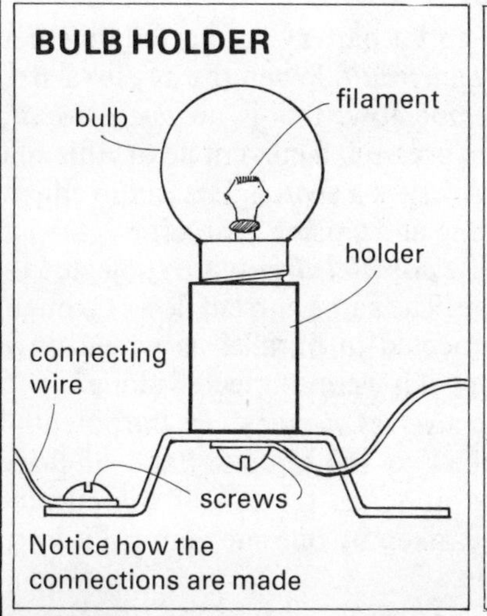
BULB HOLDER
bulb
filament
holder
connecting wire
screws
Notice how the connections are made

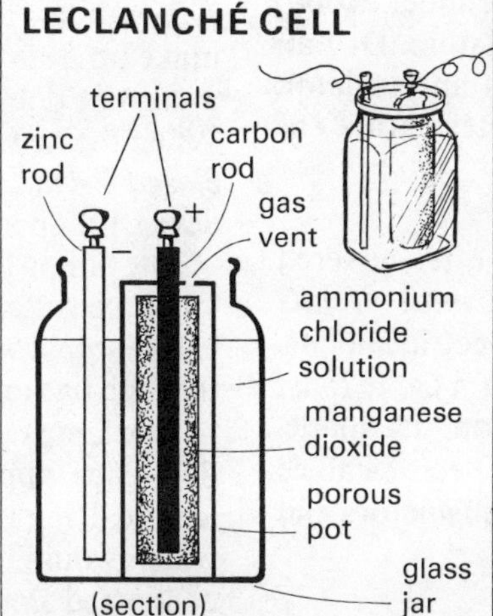
LECLANCHÉ CELL
terminals
zinc rod
carbon rod
+
−
gas vent
ammonium chloride solution
manganese dioxide
porous pot
glass jar
(section)

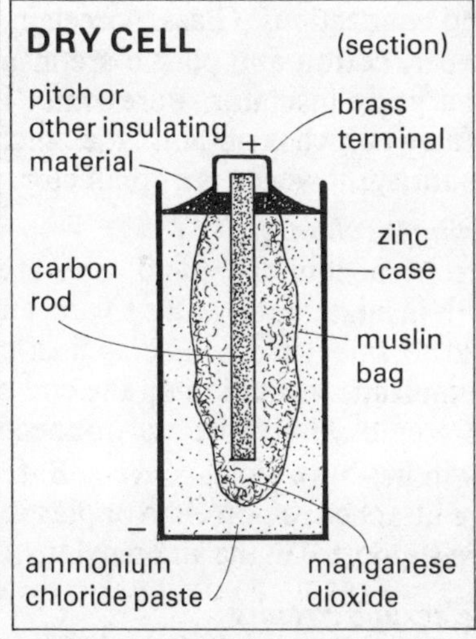
DRY CELL
(section)
pitch or other insulating material
brass terminal
+
zinc case
carbon rod
muslin bag
−
ammonium chloride paste
manganese dioxide

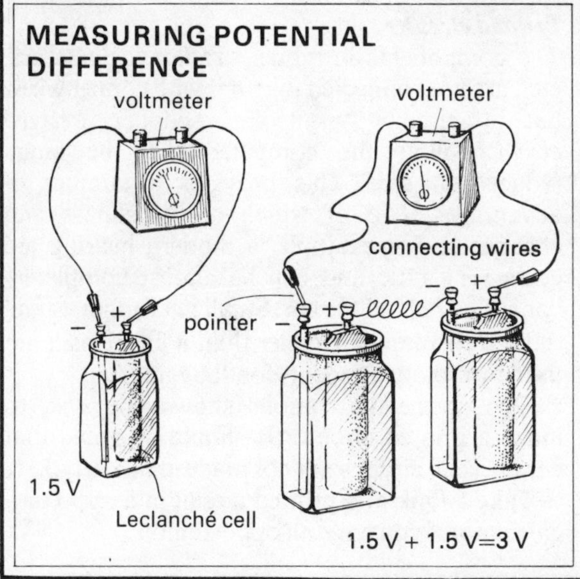
MEASURING POTENTIAL DIFFERENCE
voltmeter
voltmeter
connecting wires
pointer
−
+
1.5 V
Leclanché cell
1.5 V + 1.5 V=3 V

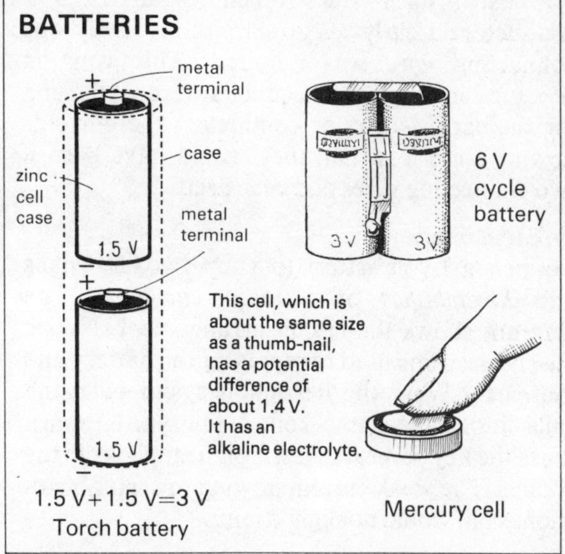
BATTERIES
+
metal terminal
case
zinc cell case
metal terminal
1.5 V
1.5 V
−
6 V cycle battery
3 V
3 V
This cell, which is about the same size as a thumb-nail, has a potential difference of about 1.4 V. It has an alkaline electrolyte.
1.5 V+1.5 V=3 V
Torch battery
Mercury cell

8 Electric Circuits

Conductors and insulators

Put a torch bulb in a holder and connect it to a 3-volt battery. The bulb lights. Cut one of the connecting wires and leave the ends apart. The bulb does not light. Use small objects, in turn, to join the ends of the severed wire. Make two lists of objects—those which allow a current to pass and those which do not.

Materials through which electricity flows are called *conductors*. Materials through which electricity does not flow are called *insulators*. Most metals are conductors. Copper is a good conductor. It is generally used for making wires and connections. Glass, porcelain, rubber, wood, paper, cotton and plastic are insulators. Dry air is a good insulator. Pure water is an insulator. Water which contains the slightest trace of impurity, however, is a conductor.

Using insulators

Electrical equipment and wires are often covered with insulating materials to prevent *short circuits* and to safeguard people against electric shocks. The plastic insulation on the end of a cable must be removed before a connection is made. Switches have plastic covers. Bare electric cables are attached to porcelain or plastic insulators and are supported in the air on pylons.

Telegraph circuits

The earth itself was sometimes used as a conductor in early *telegraph* circuits. Only one connecting wire was required. This was an economy in wire but the circuits were inefficient, for the earth is a poor conductor; the currents flowing were less than they would have been if two connecting wires had been used.

An earth return

Connect a 3-volt battery to a key, two large nails and *headphones* or a radio *earphone*. The diagram shows the circuit arrangement. Notice the special signs used to represent the battery and the earth. Wear the headphones and push the nails into some damp soil, a few yards apart. Press the key several times. You hear clicks in the phones. The weak current flowing operates headphones but would not light a lamp.

Series and parallel circuits

Connect a 3-volt battery to a key and a torch bulb. Press the key. The bulb is lit.

Connect two bulbs in *parallel* to a key and a battery. The diagram shows you how to do this. Press the key. The bulbs are lit. They shine just as brightly as did the single bulb.

Now connect two bulbs in *series* to a key and a battery. Press the key. The bulbs are dimly lit. Now connect another bulb in series. What do you notice? Use a voltmeter to measure the potential difference across each bulb.

Kinds of circuits

A key, a lamp and a battery connected together make up a *simple circuit*. When the key is raised, a current does not flow; this is an *open circuit*. When the key is pressed, a current flows; this is a *closed circuit*. A key is a switch. It is a convenient device for opening and closing a circuit.

The lamps in a *parallel circuit* are subjected to the same voltage. The same current flows through each lamp connected in parallel as would flow through one lamp if it were connected alone.

The lamps in a *series circuit* share the potential difference applied to the circuit. Each of three identical lamps in series takes $\frac{1}{3}$ of the current which would be taken by one such lamp if it were connected alone.

Printed circuits

The components in modern radios, calculators, etc., are not connected together with copper wires but by circuits "printed" in copper. *Printed circuits* allow the components to be much reduced in size. This provides a number of advantages, two of which are smallness and cheapness. For example, a modern hearing aid costs very little and can be almost completely concealed behind the ear. Small microphones and radio transmitters, smaller than a finger nail, are used by espionage agents for "bugging".

One of the illustrations shows you how to make a printed circuit. (It should be noted that commercial circuits are not made in this way.)

Take a look at a printed circuit in a radio or a calculator if you have an opportunity.

CONDUCTORS AND INSULATORS

Which are conductors?

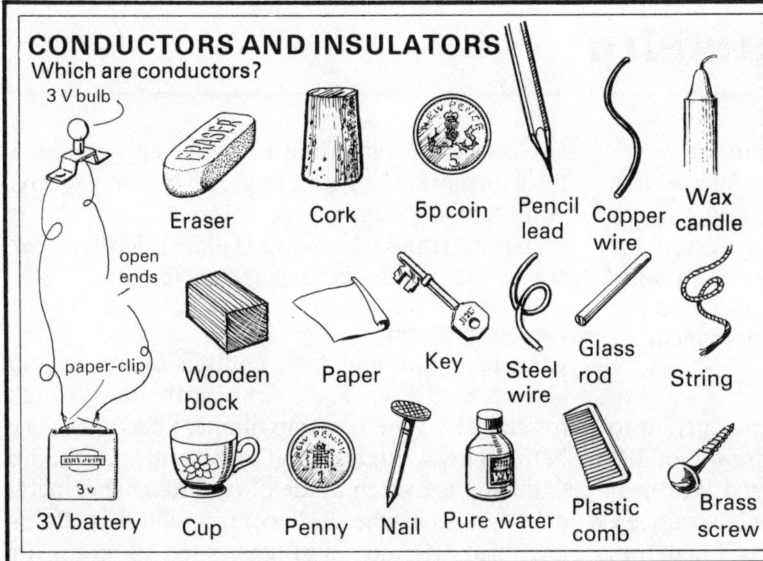

USING INSULATORS

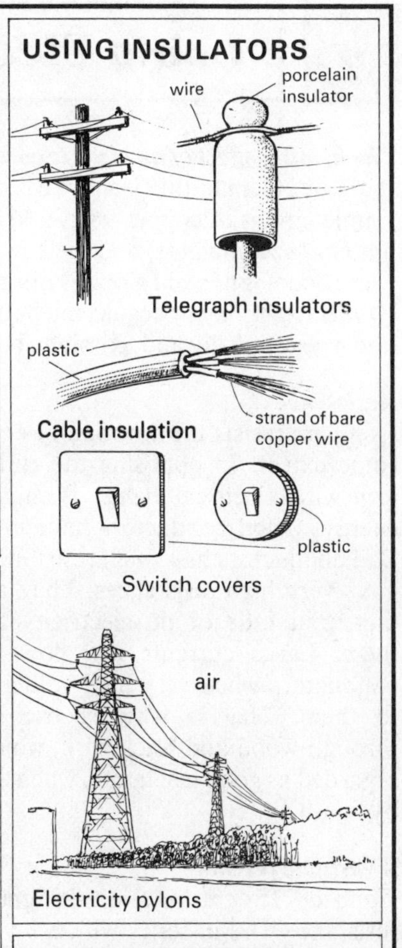

EARTH RETURN

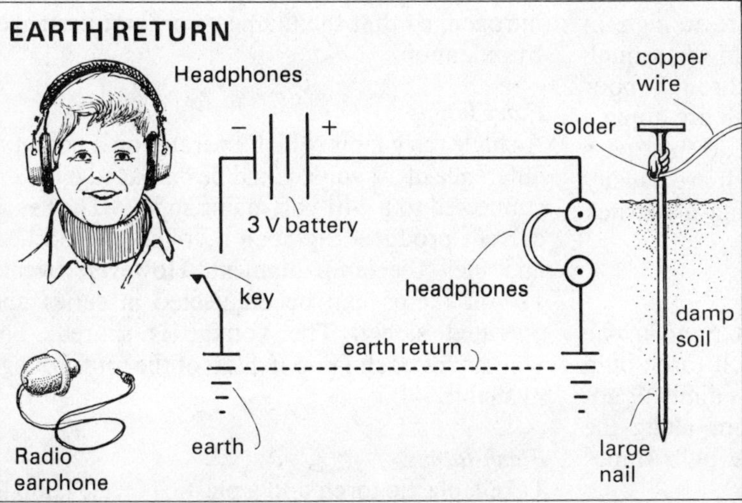

PRINTED CIRCUITS

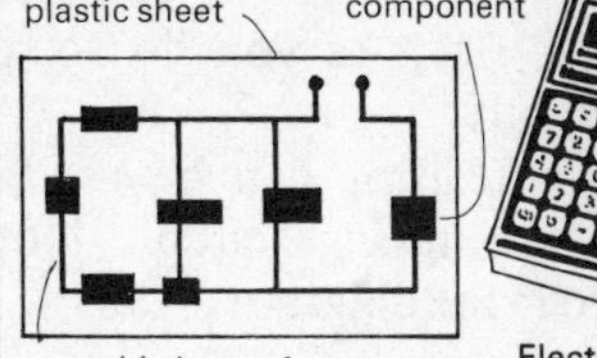

AN EXPERIMENT

1. Use a paint-brush and metallic paint (it must contain a real metal, such as copper, aluminium, etc.) to paint the circuit shown below.
2. Hold the ends of two wires from a battery on the two terminals shown.
3. Hold the ends of two wires from a voltmeter on the other two terminals shown The movement of the pointer shows that an electric current is flowing.

cardboard or wood

terminal

painted strip of metal

TELEGRAPH CIRCUITS

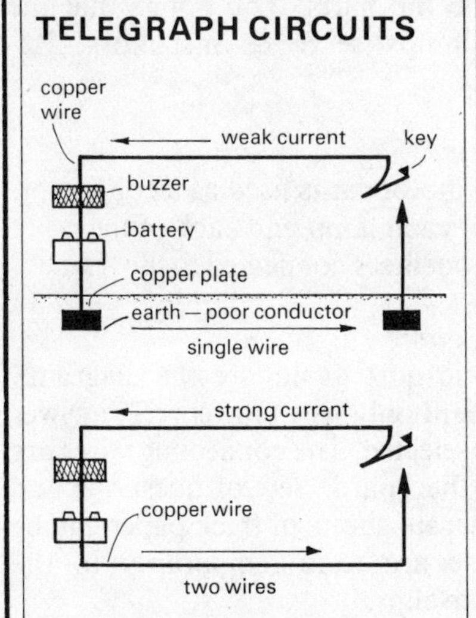

CIRCUITS

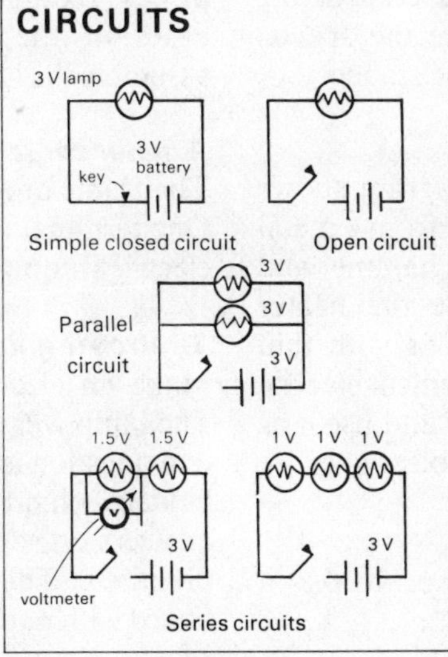

9 Using Electricity

The heating effect of an electric current
If about 10 cm of thin iron wire is connected to a simple circuit of copper wire, a key and a 6-volt battery (accumulator) and the key is pressed, the wire becomes hot and glows. *Your teacher should do this experiment* because the battery could run down very quickly and, possibly, be damaged.

Resistance
Iron wire resists the flow of an electric current to some extent. In opposing the *resistance* of the iron wire, electrical energy is changed into heat energy. Good conductors have low resistances. Bad conductors have high resistances. Insulators have very high resistances. They are so high, in fact, that little or no electricity flows through them. Large currents *will* flow through poor conductors when very high voltages are applied to them. That is why electric currents pass through wood and brickwork, which are usually regarded as good insulators, when they are struck by lightning.

A variable resistor
Connect 15 cm of No. 36 gauge *nichrome* resistance wire in series with a 3-volt torch bulb and a 3-volt battery. The bulb shines dimly or not at all. Move one of the connections along the nichrome wire. As you do so, the bulb shines more brightly.

Such a resistor can be used to control the current flowing in a circuit. Notice the special sign used in the diagram to denote resistance.

A model electric fire
Make a model electric fire. The diagram shows you how to do this. Cut the reflector from a metal can with a pair of old scissors. Use a hammer and a nail to make the holes to take the heater connections. Insulate the connections with thin glass tubes. Cut off 1/20 of a replacement electric fire *element* (this can be purchased) and use it as the heater. Connect the fire to a 12-volt supply.

An electric lamp
Examine a 240-volt electric lamp. Make a labelled drawing.

Connect the lamp to a 120-volt supply (from a transformer). The filament glows but only dimly. A potential difference of at least 200 volts is required to make the filament glow brightly. *Your teacher should do this experiment for you.*

Lamps and fires
Electric lamps and fires contain filaments and elements. These are wires with fairly high resistances. When a lamp filament becomes very hot, it glows brightly and gives out light. Lamp filaments are often made of *tungsten.* This metal does not melt when it becomes white-hot. Bulbs are filled with an inert gas, such as argon or nitrogen, so that the filaments are not destroyed by oxidation.

Fairy lamps
A single fairy lamp which operates on a potential difference of 12 volts would be damaged if it were connected to a 240-volt mains supply. The heavy current produced by such a voltage would heat and melt the lamp filament. However, twenty 12-volt lamps can be connected in series and operated safely. The voltage is shared. The voltage for each lamp is 1/20 of the total voltage available.

Flash-lamps
Take a plastic torch and a plastic bicycle lamp to pieces. Examine the parts. You notice that the cases of the lamps serve as insulators. Re-assemble the lamps.

A motor car frame
The frame of a motor car is used as a conductor. One terminal of each lamp and each of the other electrical components is connected to the frame.

An electric quiz game
Make an electric quiz game (see the diagram). The lamp will light only when the correct answer to a question is selected. The connecting wires are hidden behind the board. Sets of questions and answers on separate sheets of thick paper can be prepared. These are fixed temporarily to the board with paper-clips.

THE HEATING EFFECT OF A CURRENT

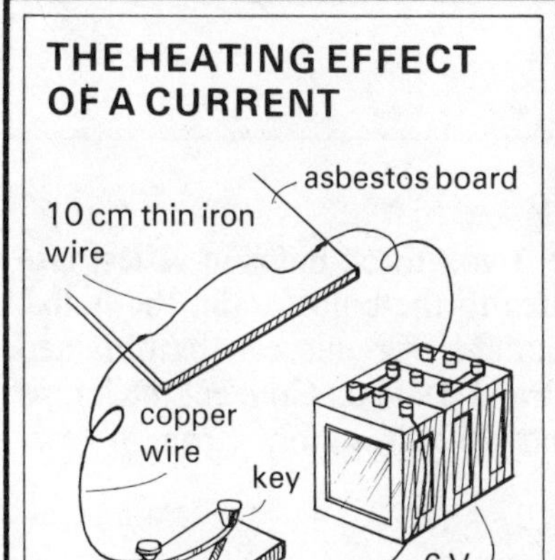

VARIABLE RESISTOR

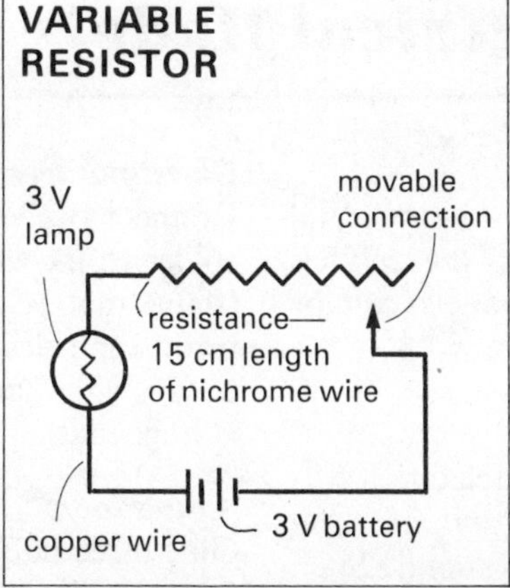

MODEL ELECTRIC FIRE

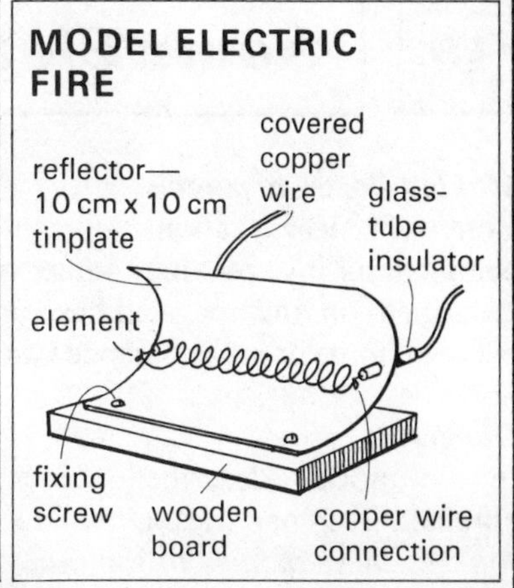

ELECTRIC LAMP

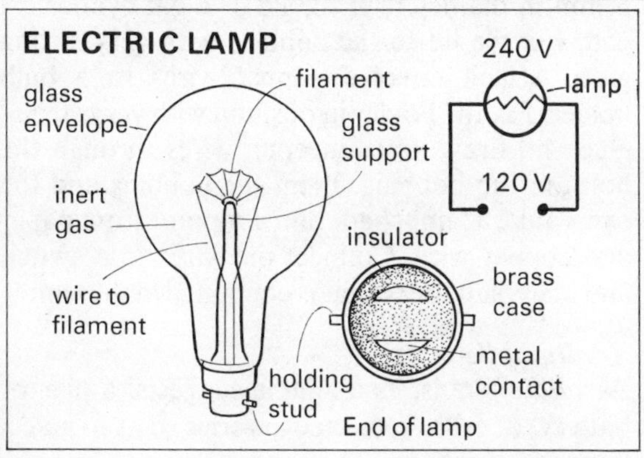

TORCH

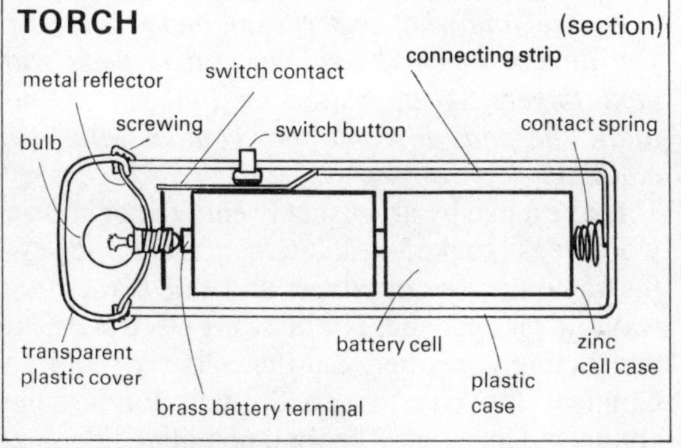

MOTOR CAR FRAME

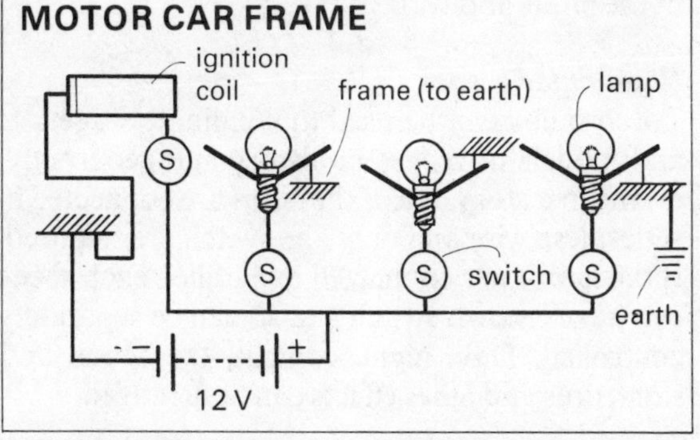

FAIRY LAMPS

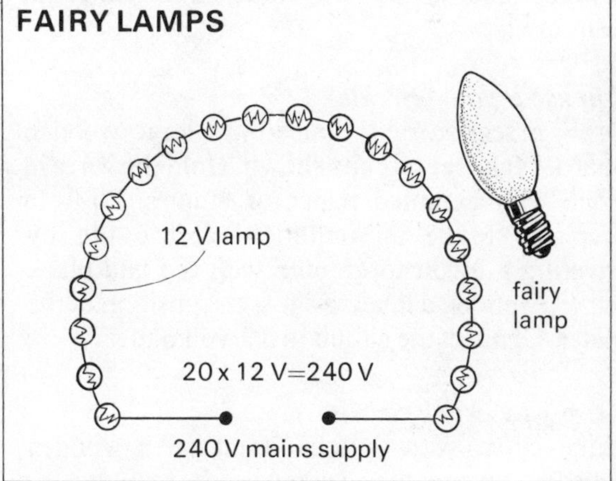

ELECTRIC QUIZ GAME

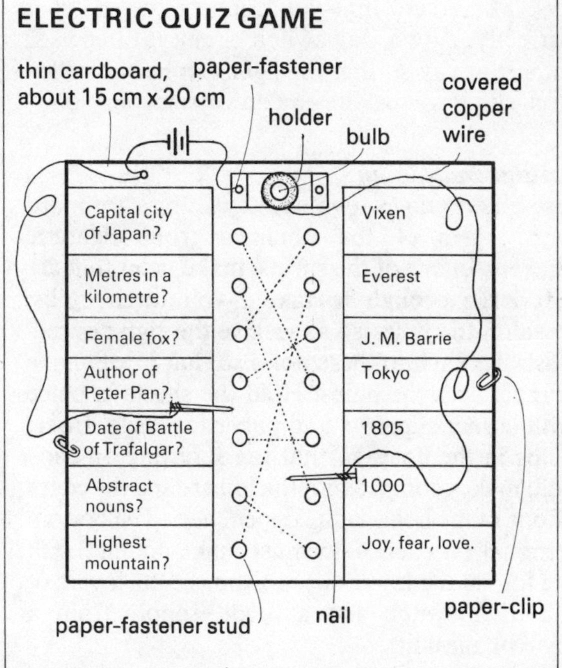

10 Some Electrical Models

Making electrical models

You will learn about correct wiring and connections by making some of the models described on this page. These models can be made at home or in the Science Club.

Danger

In the models described below, electricity is obtained from batteries. *On no account must you connect these models to the mains supply.* Such action would damage the models and you could be injured.

A house-lighting circuit

Draw a section of a house on a large wooden board. Make the circuits shown. Hold the wires in place with gummed paper or staples. Draw a fireplace. Make an imitation electric fire by covering a 3-volt torch bulb with red and black paper. Gum on a label to show the position of the meter. Connect the circuit to a 3-volt battery.

A two-way switch circuit

Make a two-way switch circuit on a wooden board. Two two-way switches are required for this. Make sure that you connect the switches correctly. A two-way switch circuit is often used on a stairway so that the light can be controlled from either an upstairs or a downstairs position.

An illuminated sign

Use glue to cover two sheets of thin cardboard with layers of foil obtained from cigarette packets. In one of the sheets, make holes that are just large enough to take 3-volt torch bulbs. Position the holes so as to give the sign desired. Push the foil into these holes so that it will make contact with the bulbs. Hold the sheets in place with corner supports and paper fasteners. Insert bulbs in the holes so that the screwing on each bulb makes contact with the foil around the edges of one of the holes in the upper sheet. The bottom terminal on each bulb must make contact with the foil on the lower sheet. Connect the layers of foil to a switch and a 3-volt supply from a low-voltage unit.

Christmas tree lights

Connect twelve 3-volt torch bulbs in series. Use solder to fix wires to the bulbs. Paint the bulbs. (Paint that is suitable for this can be obtained from some electrical dealers.) Connect the lamps to a switch and 36-volt supply (from a low-voltage unit).

A table lamp

Glue three bobbins together. Make a hole, about 5 mm in diameter, in the lid of a flat cylindrical can. Fix the lid to the bobbins with very strong glue. Attach covered copper wires to a bulb holder. Fix the holder in position with very strong glue and draw its connecting wires through the holes in the bobbins. Paint the bobbins and the can. Make a lampshade and a support from paper and copper wire. Connect the wires to a switch and a 3-volt battery that is contained in the can.

A voltaic pile

A *voltaic pile* is, as its name suggests, a pile, or battery, of cells connected in series so as to give a fairly large voltage.

In the following experiment make sure that you do not touch the sulphuric acid pads with your fingers. Also, make sure that you do not allow the pads to come into contact with your clothing.

Make a pile by arranging twenty felt or cotton-wool pads, soaked in dilute sulphuric acid, and the same number of copper and zinc discs in the way shown opposite. The discs are electrodes. No connecting wires between the cells are required. Connect the pile to seven 3-volt torch bulbs connected in series. The total potential difference of the pile is about 20 volts (20×1).

Illuminated scenes

Torch bulbs can be used to illuminate stage sets and models of scenes in history and geography, events in a story, etc. If the bulbs are connected in series, less wire and only one switch are required. If the lamps are connected in parallel, each scene can have its own switch and so can be separately controlled. Day, night, twilight, the Moon and stars, fires and other effects can be arranged.

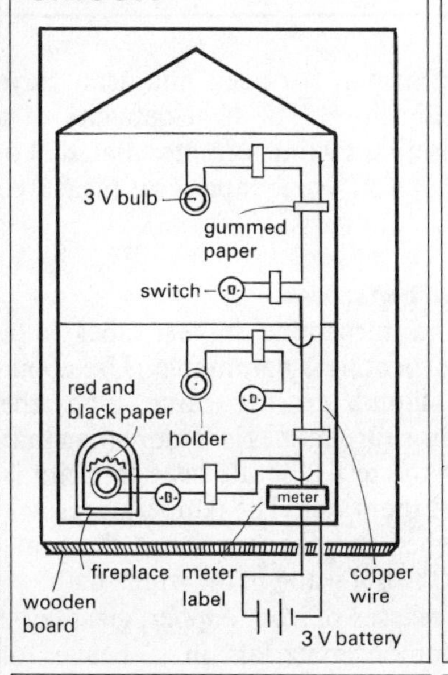
HOUSE-LIGHTING CIRCUIT
3 V bulb
gummed paper
switch
red and black paper
holder
meter
fireplace
meter label
copper wire
wooden board
3 V battery

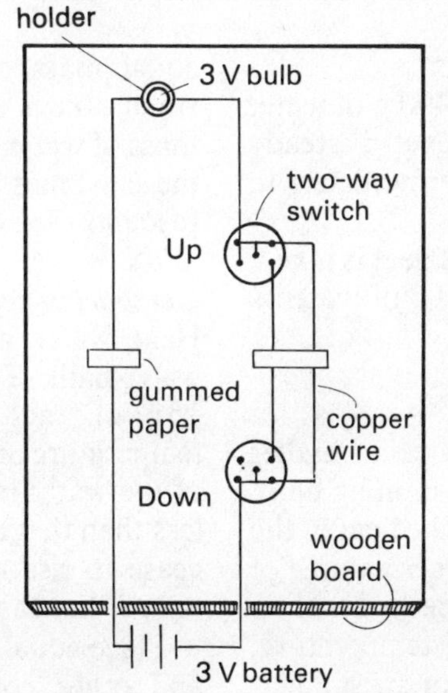
TWO-WAY CIRCUIT
holder
3 V bulb
two-way switch
Up
gummed paper
copper wire
Down
wooden board
3 V battery

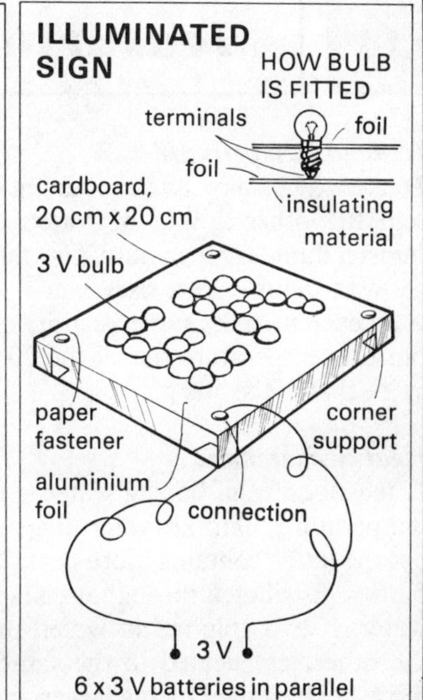
ILLUMINATED SIGN
HOW BULB IS FITTED
terminals
foil
foil
insulating material
cardboard, 20 cm x 20 cm
3 V bulb
paper fastener
corner support
aluminium foil
connection
3 V
6 x 3 V batteries in parallel

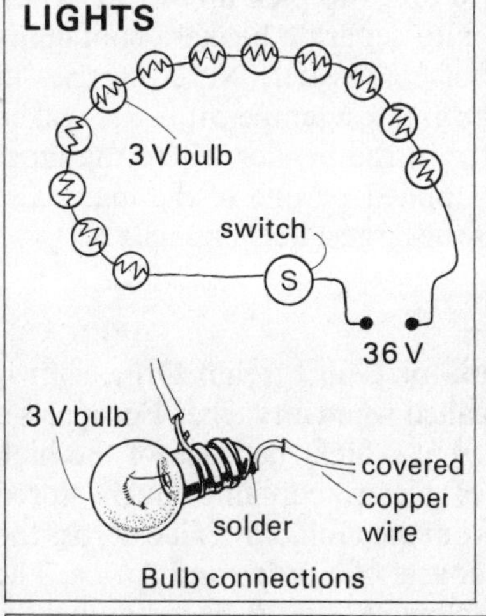
CHRISTMAS TREE LIGHTS
3 V bulb
switch
S
36 V
3 V bulb
covered copper wire
solder
Bulb connections

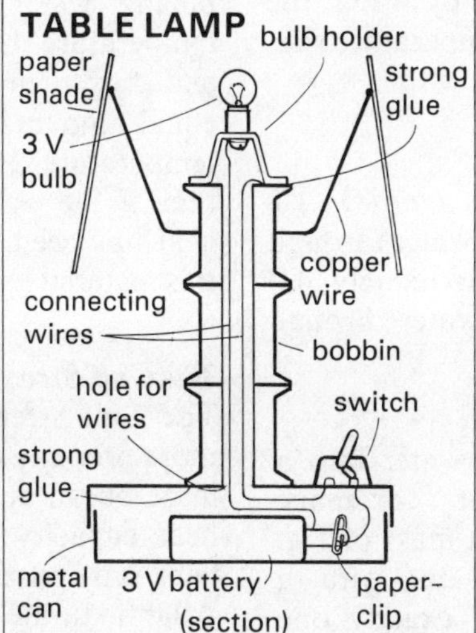
TABLE LAMP
bulb holder
paper shade
strong glue
3 V bulb
copper wire
connecting wires
bobbin
hole for wires
switch
strong glue
metal can
3 V battery
(section)
paper-clip

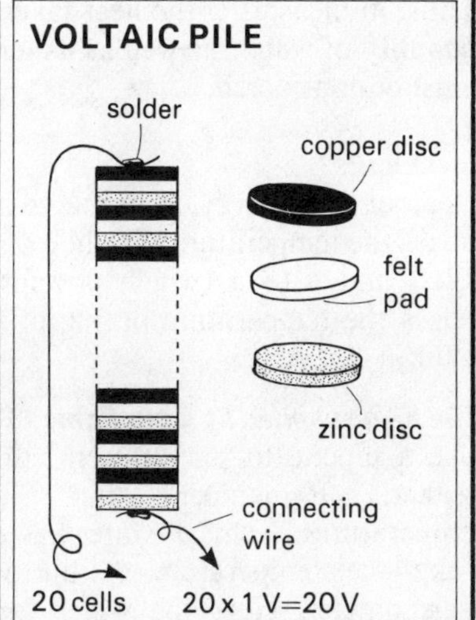
VOLTAIC PILE
solder
copper disc
felt pad
zinc disc
connecting wire
20 cells
20 x 1 V=20 V

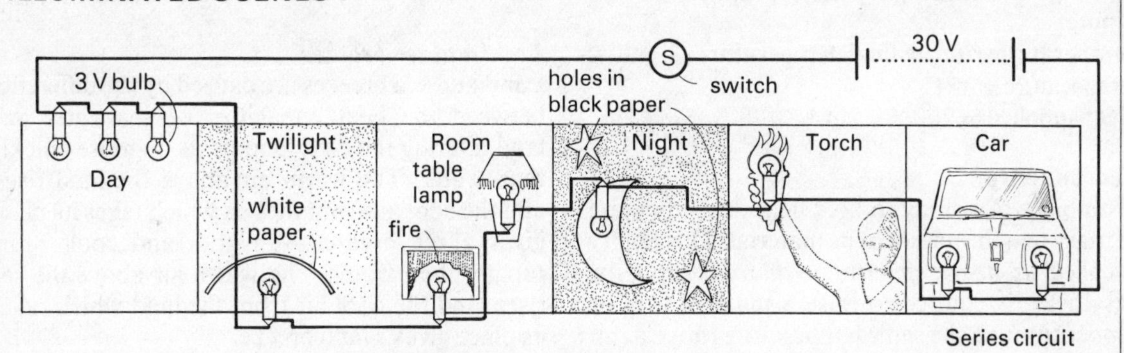
ILLUMINATED SCENES
30 V
S
switch
3 V bulb
holes in black paper
Day
Twilight
white paper
Room
table lamp
fire
Night
Torch
Car
Series circuit

11 Measuring Heat

Heat and temperature
Heat two beakers, one containing 400 g of water and the other 200 g of water, over a steady bunsen flame for 1 minute. Use thermometers to show the temperature rises.

In each case the same quantity of heat is taken, but the temperature rise of the 200 g of water is twice that of the 400 g.

Heat measurement
A teaspoonful of boiling water is at a far higher temperature than the water in a swimming bath, yet the latter contains more heat. The heat in the bath is distributed throughout a larger volume of water. If two volumes of water, one larger than the other, are heated to the same temperature, more heat is taken by the larger than the smaller. Thus, in measuring the heat taken by water, the quantity of water, as well as its temperature rise, must be considered.

Heat units
Heat, or heat energy, is measured in *joules* (J). 1 J raises the temperature of 0.24 g of water through 1°C, and 4.18 J (which is approximately 4J) raises the temperature of 1 g of water through 1°C.

The heat supplied by a gas flame
Use a pipette to put 200 cm^3 of water into a beaker. Remember that, at ordinary temperatures, 1 cm^3 of water has a mass of 1 g. Take the temperature of the water with a thermometer. Heat the water for exactly one minute and then take its temperature again. Calculate the heat, in joules, supplied in one minute.
Temperature rise = final temperature − initial temperature = x°C
Heat supplied = $200 \times x \times 4$ J

Heat capacity
The ability of a body to hold heat depends upon its mass and the nature of its material. This ability is called its *heat capacity,* or *thermal capacity.* Less heat is required to raise a mass of iron or copper through the same temperature range as an equal mass of water. Copper and iron have smaller heat capacities. The heat capacity of a mass of water is about 9 times greater than that of the same mass of iron. Heat capacity is measured in *joules per °Celsius* (J/°C).

Comparing heat capacities
Heat water and turpentine in test-tubes in a water-bath. (Turpentine is flammable.) Use about 15 g of each liquid. In the same time, the temperature of the turpentine rises more than that of the water in the test-tube; its heat capacity is less than that of the water. The temperatures will cease to rise when they have reached the same temperature as the hot water in the water-bath.

Place equal masses of iron, copper, glass, zinc and other common materials in a beaker of boiling water so that all acquire the same temperature. Use tongs to pick up the materials, and drop them into separate beakers containing equal amounts of cold water. Note the rises in temperature. The greatest temperature rise will be that of the water in the beaker where the most heat has been supplied by one of the materials; this material has the largest heat capacity.

The Gulf Stream
The Gulf Stream, or Gulf Stream Drift, as it is more properly called when it reaches Europe, is a large ocean current which, because of the high heat capacity of water, contains much stored heat. It travels from the Gulf of Mexico across the Atlantic to the coasts of Europe and Africa. The westerly winds blowing over it become warmer and influence the climate of Britain.

Land and sea breezes
Land and sea breezes are caused by the difference between the heat capacities of sea-water and land. During the day, land heats up more quickly than water. The warm air above the land rises, and the cool air from the sea which takes its place gives a sea breeze. At night, land cools more rapidly than water. The warm air above the sea rises, and the cool air from the land which takes its place gives a land breeze.

HEAT AND TEMPERATURE

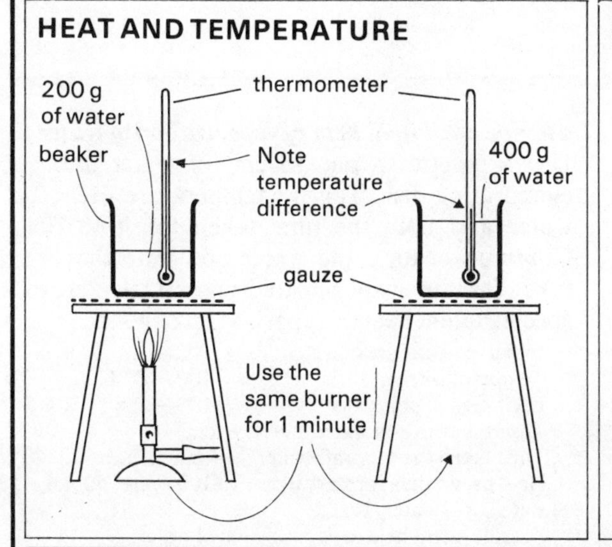

HEAT SUPPLIED BY A GAS FLAME

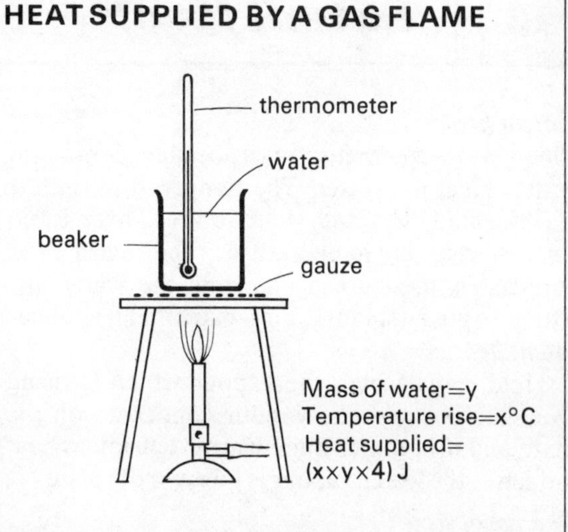

HEAT CAPACITY

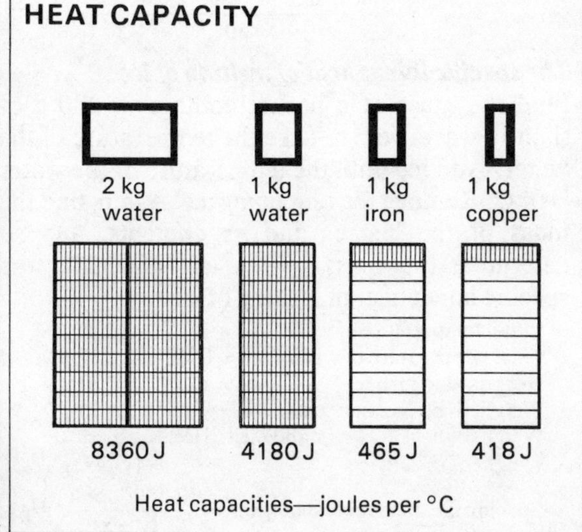

COMPARING HEAT CAPACITIES

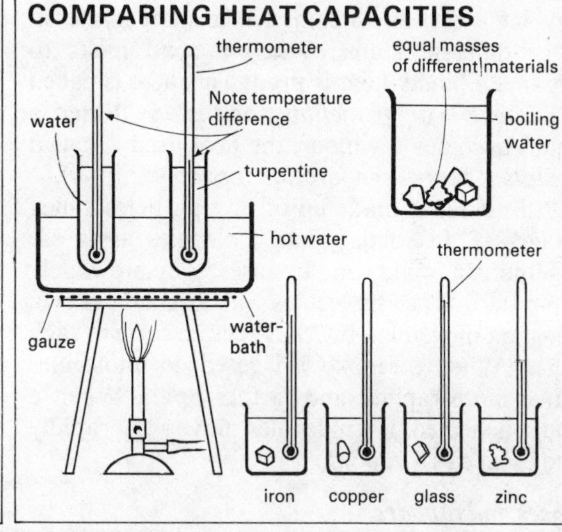

GULF STREAM

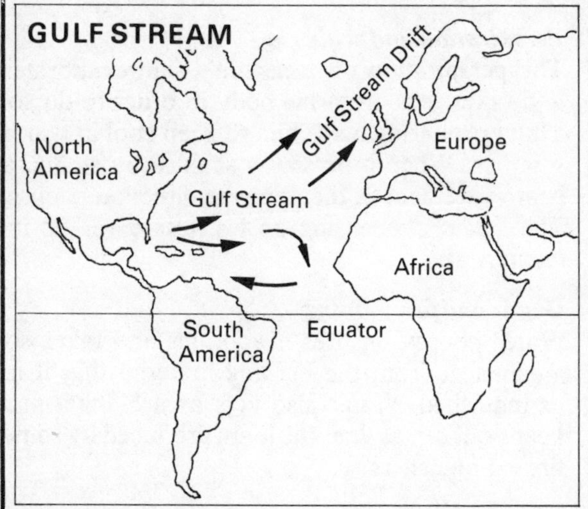

LAND AND SEA BREEZES

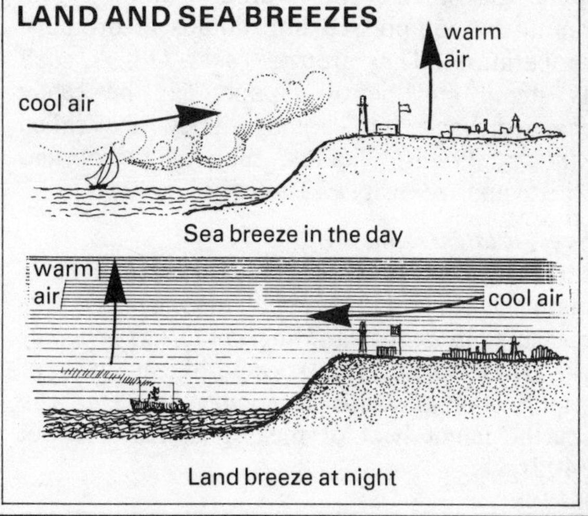

12 Latent Heat

Latent heat

Place a thermometer in a beaker containing water. Heat the water. The temperature rises to 100°C, and then the water boils. There is no further rise in temperature. The extra heat supplied is used up in changing the water into water vapour (steam). This extra heat is called *latent heat.*

Hold a plate over the vapour which is rising from the beaker. The vapour condenses on the plate and it becomes quite warm. The latent heat hidden in the water vapour is given to the plate.

Change of state

Matter exists in three states—*solid, liquid* and *gas*. Ice, water and steam are the same matter in three different states. When a solid melts to become a liquid, heat is used. This heat is called the *latent heat of melting* (or *fusion*). When a liquid becomes a vapour, the heat used is called the *latent heat of evaporation* (or *vaporization*).

All matter is made up of tiny particles called molecules. In solids, these molecules are close together. In liquids and in gases, they are widely separated. The molecules in a solid are in constant movement but they remain close to each other. When the solid is heated, its molecules move more rapidly and farther apart. When a liquid is heated, its molecules move very rapidly and escape as a vapour.

Gases and vapours

Some gases, such as hydrogen and oxygen, cannot be compressed into liquids at ordinary temperatures. They are *real gases*. Others, such as steam and petrol vapour, can be easily changed back into liquids. They are called *vapours*. *Volatile* liquids, such as petrol and ether, vaporize easily and quickly.

Specific latent heat

The heat required to change the state of 1 kg of matter is called its *specific latent heat;* it is measured in *joules per kilogram* (J/kg) or *joules per gram* (J/g). The specific latent heat of vaporization of water is about 2200 J/g. The specific latent heat of melting of ice is about 340 J/g.

The specific latent heat of vaporization of water

Use a pipette to put 50 cm³ of water into an evaporating dish. Take its temperature. Heat the water and note the time taken for it to boil. Continue heating the water and note the time taken for it to evaporate to dryness. Calculate the specific latent heat of vaporization of water.

Initial temperature of water $= z\ °C$
∴ Temperature rise $= (100 - z)\ °C$
∴ Heat used to boil water $= (100 - z) \times 50 \times 4$ J
Time taken to boil water $= x$ min
Time taken to evaporate water $= y$ min
∴ Heat used to evaporate water $= (100 - z) \times 50 \times 4 \times \frac{y}{x}$ J
But 50 g of water is used.
∴ Specific latent heat of vaporization of water =

$$\frac{\text{heat used to evaporate water}}{50}\ \text{J/g}$$

The specific latent heat of melting of ice

Find the mass of a beaker containing 200 g of slightly warm water. Take the temperature of the water. Add ice until the temperature of the water is 0°C. Remove the remaining ice. Again, find the mass of the beaker and its contents, and so determine the mass of the ice. Calculate the specific latent heat of melting of ice.

Initial temperature of water $= Z\ °C$
Final temperature of water $= 0°C$
∴ Heat used to melt ice $= (Z - 0) \times 200 \times 4$ J
Weight of beaker and water $= x$ g
Weight of beaker, water and melted ice $= y$ g
∴ Weight of melted ice $= (y - x)$ g

∴ Specific latent heat of melting ice $= \dfrac{Z \times 200 \times 4}{(y - x)}$ J/g

Perspiration and scalds

The perspiration on a person's skin evaporates and takes heat from his body in order to do so. Thus, perspiring helps him to keep cool in warm weather. When a person is scalded by steam, a heat movement in the opposite direction occurs. The steam condensing on his flesh gives up its latent heat.

Water and fire-fighting

Water poured on fires evaporates and takes so much heat from the burning material that it is extinguished. Water also acts as a blanket and keeps out air, as does the foam produced by some fire extinguishers.

LATENT HEAT

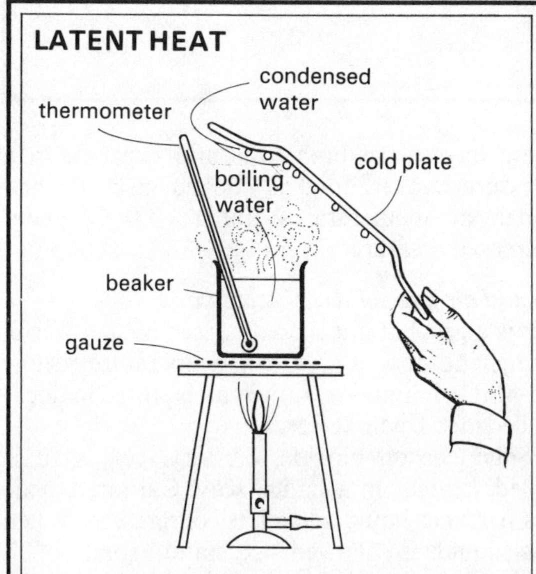

CHANGE OF STATE

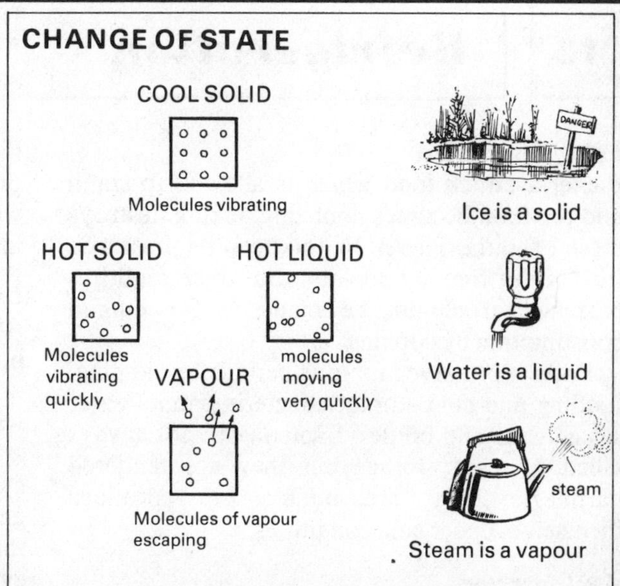

SPECIFIC LATENT HEAT OF VAPORIZATION OF WATER

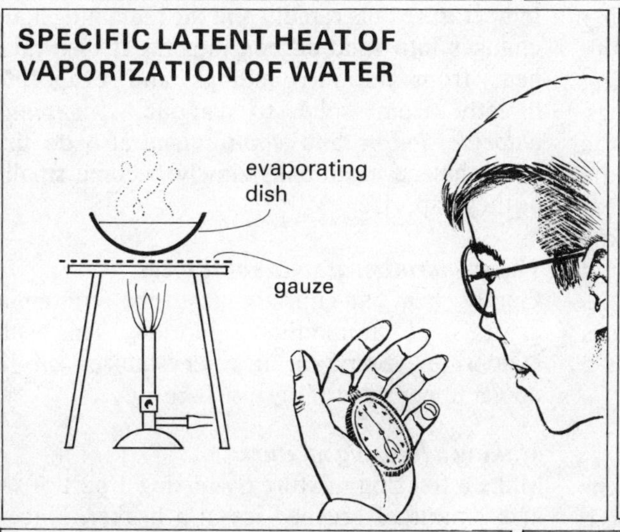

SPECIFIC LATENT HEAT OF MELTING OF ICE

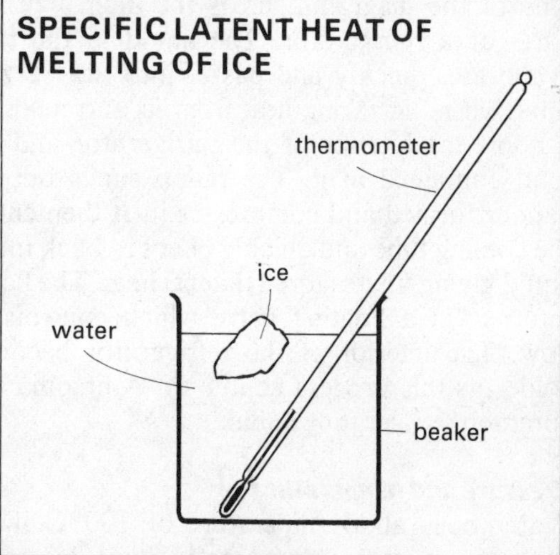

PERSPIRATION AND SCALDS

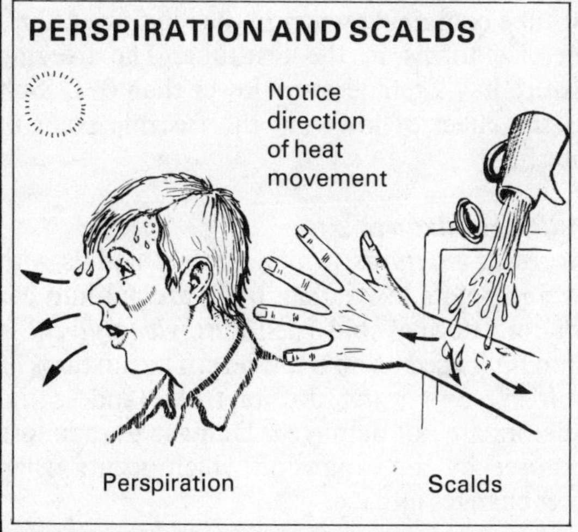

WATER AND FIRE-FIGHTING

Evaporating water uses latent heat

13 Refrigeration

Food preservation

Bacteria attack food which is allowed to stand and the food becomes tainted. Cooking destroys these bacteria. Therefore, food can be preserved for a short time by cooking. But once food has putrefied, it cannot be made fit for human consumption by cooking.

Food is preserved for long periods by canning, bottling and cold storage. Bacteria cannot enter sealed cans and bottles. Bacteria are not always killed by cold storage but they are rendered harmless. They are unable to reproduce themselves under cold conditions.

A refrigerator

One of the diagrams shows the main working parts of a refrigerator. The liquid in the tank evaporates quickly and passes into the freezing tube, where, in taking heat from its surroundings, it cools the interior of the refrigerator and the food contained in it. The pump sucks out the vapour formed and compresses it. It then enters the cooling tube and quickly changes back into a liquid giving up its stored (latent) heat. The liquid returns via a floating valve which controls its flow. The interior of the refrigerator becomes colder as this process continues. Ammonia is a common refrigerating liquid.

Pressure and temperature

Water boils at a temperature of 100°C if the pressure on it is atmospheric, that is, about 100 000 pascals (100 000 N/m^2). On a mountain top, where the atmospheric pressure is less than it is at sea-level, water would boil at a temperature lower than 100°C. Water vapour molecules escape very easily from the surface of heated water if the pressure of the air above it is reduced.

Liquids boil at low temperatures if they are heated in vessels in which the pressure has been reduced. A pump draws vapour out of a refrigerator tank, the pressure within it is reduced, and the refrigerating liquid in the tank vaporizes even at very low temperatures.

Pressure cookers

Pressure cookers, which are in fairly common use these days, save both time and expense. In a pressure cooker, food is cooked in steam and water at temperatures above 100°C under increased pressure.

Liquid air and solid carbon dioxide

Air is a gas, but if it is compressed by a powerful pump and cooled to a low temperature (below −140°C), it liquefies. Liquid air is stored in thick-walled metal cylinders.

Solid carbon dioxide, or "dry ice", as it is called, is made in a similar way. Carbon dioxide gas becomes liquid when it is compressed. When this liquid is allowed to stand, some of it evaporates. Heat in it is used in evaporation; its temperature falls rapidly and the remaining liquid changes into solid carbon dioxide. It then takes heat from its surroundings and evaporates directly from solid to vapour. *Ammonium chloride, iodine* and *naphthalene* also do this. Naphthalene moth-balls slowly become smaller as they vaporize.

The evaporation of solid substances

Gently heat an upright test-tube containing crystals of ammonium chloride. This solid evaporates. Some of it re-crystallizes on the cooler upper sides of the test-tube.

Making a freezing mixture

Make a freezing mixture by mixing 1 part of salt and 2 parts of crushed ice in a beaker. Place a test-tube containing water in the mixture. After a time, ice forms in the test-tube. The freezing mixture has a temperature lower than 0°C. Salt has the effect of lowering the freezing-point of water.

Antifreeze mixtures

Antifreeze mixtures, which contain liquids with freezing-points lower than 0°C, are put into car radiators during cold spells. *Ethylene glycol* is commonly used as an antifreeze. The mixtures of antifreeze and water do not freeze and so the radiators are not damaged. Damage to radiators is caused by the expansion which occurs when water changes into ice.

FOOD PRESERVATION

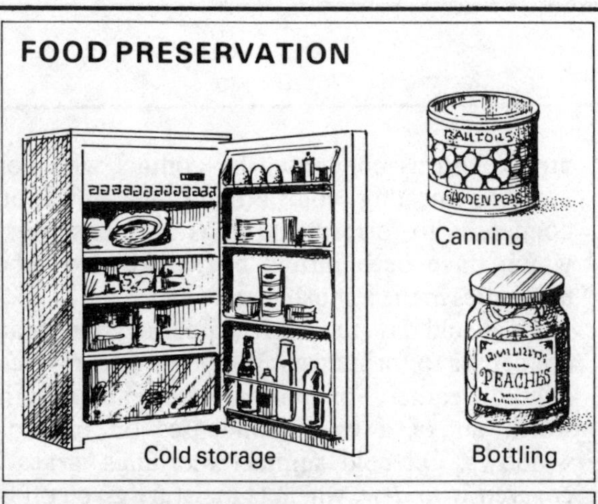

Cold storage

Canning

Bottling

PRESSURE COOKER

The water in the cooker is under increased pressure, and so it boils at a temperature above 100°C.

EVAPORATION OF SOLID SUBSTANCES

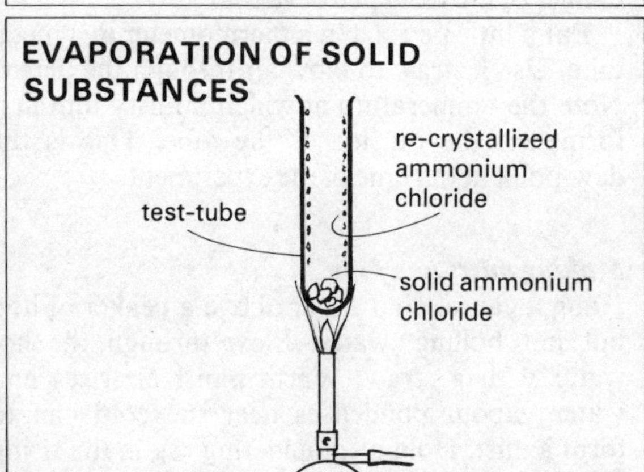

MAKING A FREEZING MIXTURE

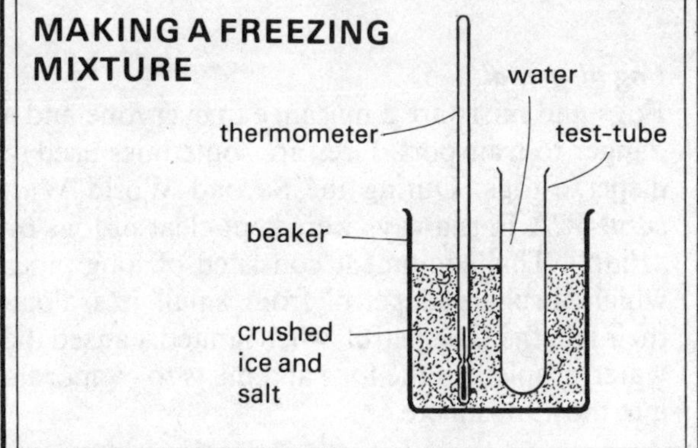

REFRIGERATOR

(diagrammatic)

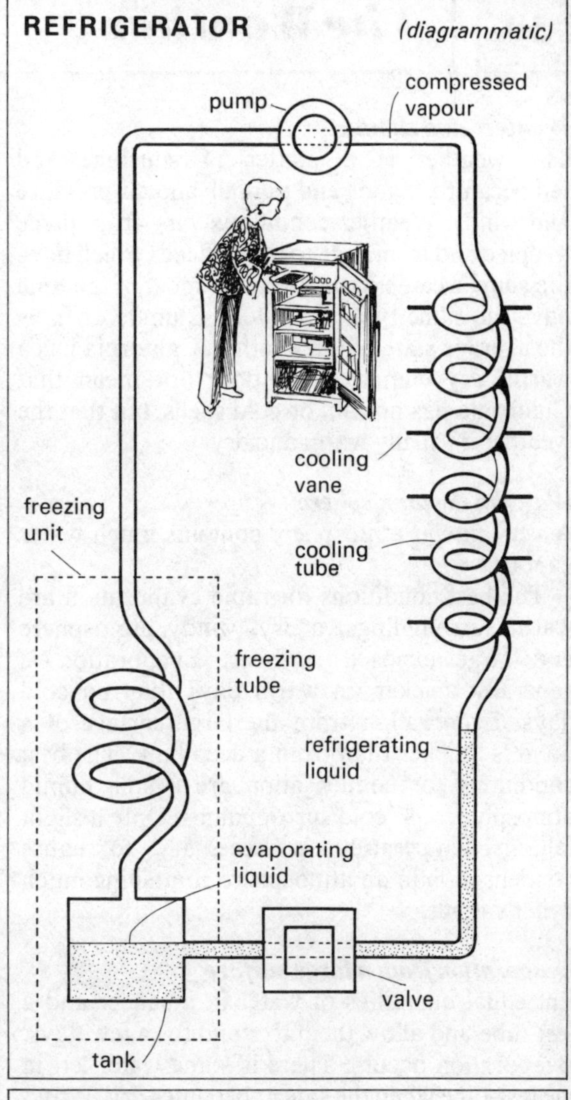

ANTIFREEZE MIXTURES

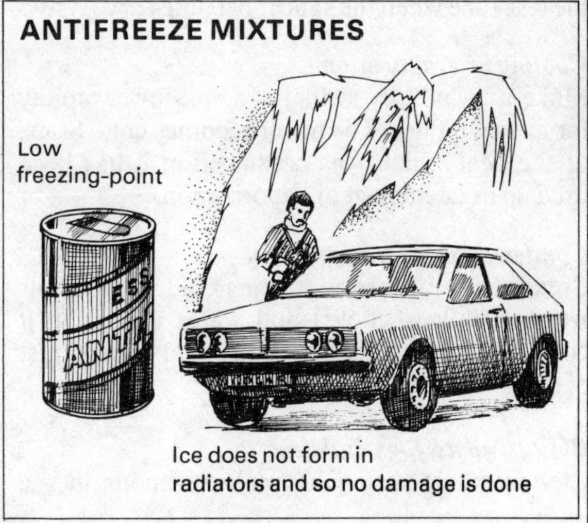

Ice does not form in radiators and so no damage is done

14 The Weather

Weather and climate

The weather is a matter of sunshine and temperature, clouds and rainfall, and air pressure and winds. Weather conditions vary from place to place and from time to time. Places which have the same weather conditions for most of the time have the same type of *climate*. Climate, then, is the average state of the weather. California has a warm, dry climate. This does not mean that California has no rain or cold spells, but that the weather is usually warm and dry.

Water in the atmosphere

A very *humid* atmosphere contains much water vapour.

The best conditions for rapid evaporation are warm surroundings, a dry, windy atmosphere and large exposed surfaces. Evaporation is generally quicker on warm days than on cold days. Evaporation from the large surface of a pond is quicker than from a deep tank. The best conditions for condensation are a still, humid atmosphere and cold surroundings. Only a slight fall in temperature is necessary to cause condensation in an atmosphere containing much water vapour.

Evaporation from a large surface

Put equal quantities of water in a saucer and a test-tube and allow them to stand for a few days. Evaporation occurs. There is some water left in the test-tube when the saucer becomes dry. Why?

Cooling by evaporation

Make a "wind" by swinging a wet towel rapidly for a few minutes. The towel becomes cold. Some of the heat which was contained in it has been used up as latent heat of vaporization.

Condensation of water vapour

Breathe on a beaker containing cold water. Your warm breath is cooled and water vapour in it condenses on the outside of the beaker as a mist of small water droplets.

Clouds, mists, fogs and dew

Clouds are formed when water vapour in the atmosphere is condensed by contact with cold air. On meeting cold ground, water vapour condenses to form mists. Fogs are dense mists which have been dirtied by smoke and other impurities in the atmosphere.

On a cold day, the water vapour in your breath condenses to form mists. Windows "steam over" in cold weather. Some of the water vapour in the warm air of a room condenses on the cold windows. On cold summer mornings, grass is covered with *dew,* which is moisture given off by plants; it cannot evaporate quickly in cold air. The temperature at which water vapour condenses, or dew evaporates, is known as the *dew-point,* and, of course, it varies with the humidity and the temperature of the atmosphere.

Finding the dew-point

Make sure that the bunsen burners are out when you carry out this experiment because of the danger of fire from petrol vapour.

Put a little petrol and a thermometer in a metal tube. Use a straw to blow air through the petrol. Note the temperature at which a misty film first forms on the outside of the tube. This is the dew-point at the time of the experiment.

Making mists and fogs

Hang a can of cold water above a beaker of hot, but not boiling, water. Blow through the hot water with a straw. Warm moist air rises and water vapour condenses near the cold can to form a mist. Hold a smouldering rag in the rising air. A fog is formed.

Fog dispersal

Fogs and mists are a nuisance to everyone and a danger to transport. Fires are sometimes used to disperse fogs. During the Second World War, some R.A.F. runways were kept clear of fogs by "Fido". This equipment consisted of long pipes which discharged petrol from small jets along their lengths. The petrol, when ignited, caused the water droplets in the fogs and mists to evaporate into the atmosphere.

WATER IN THE ATMOSPHERE

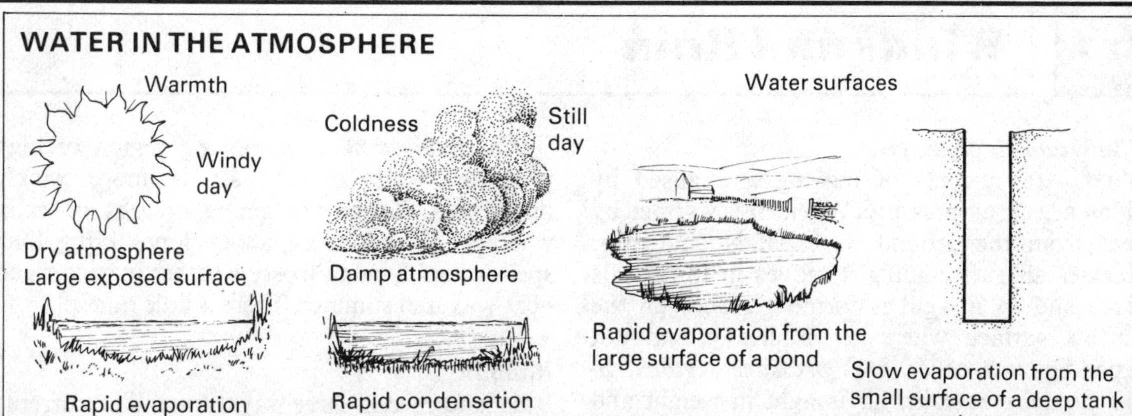

EVAPORATION FROM A LARGE SURFACE

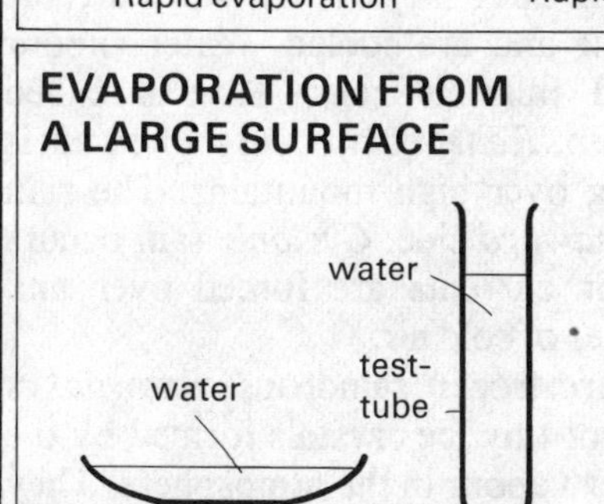

COOLING BY EVAPORATION

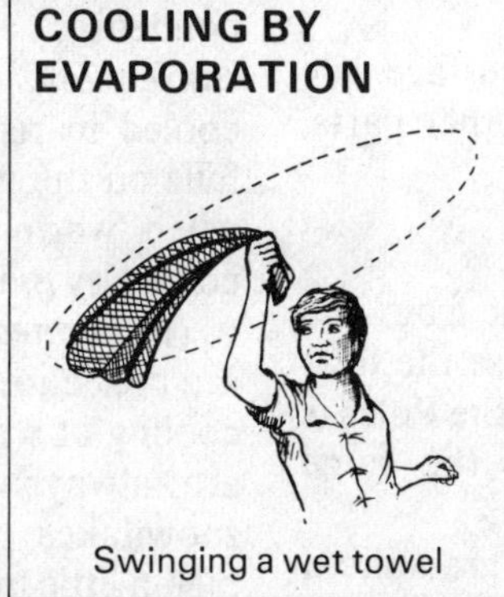

CONDENSATION OF WATER VAPOUR

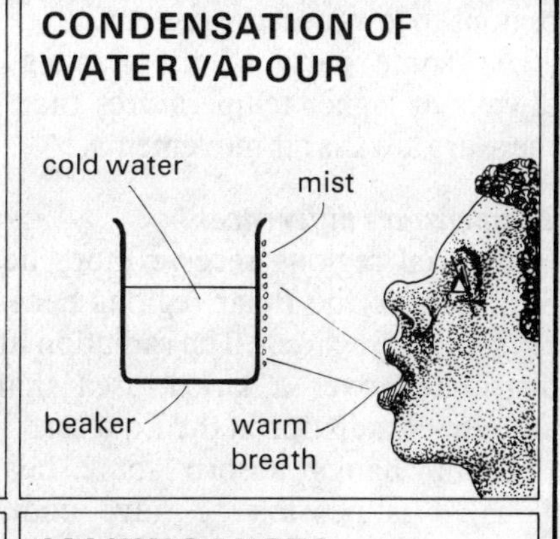

MISTS

FINDING THE DEW-POINT

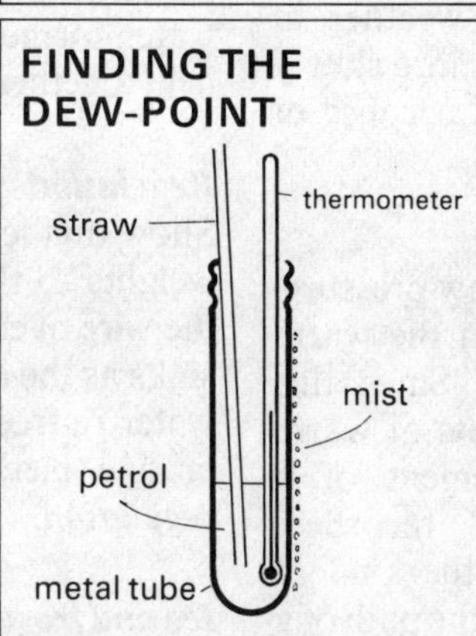

MAKING MISTS AND FOGS

FOG DISPERSAL

"Fido"

15 Winds and Rain

Winds and air pressure

Winds are currents of moving air caused by differences in air pressure. When air is warmed by heat from the ground, it expands and rises. Heavier air surrounding it moves in to take its place and so a wind is formed. Places on the Earth's surface where air is heavy and not expanding are called *high pressure regions*. In *low pressure regions*, air is light in weight and expanding. Air always moves from high pressure regions to low pressure regions.

As some parts of the Earth's surface are always at higher temperatures than other parts, there are always air movements.

Temperature differences

Equatorial regions receive more heat from the Sun than do the Polar regions because the heat rays are more direct. The radiation at the Poles is distributed over a larger area than the same amount of radiation at the Equator.

Cloud banks absorb some heat radiation. Thus, it is possible to have cool weather in tropical regions. Sea-water heats up more slowly and retains its heat longer than land because of the high heat capacity of water.

Depressions and anticyclones

A *depression* is a moving region of low pressure. Cold air from the Arctic mixes with the warm westerly winds from the Atlantic. Since this mixing is incomplete, isolated pockets of warm moist air are formed. The movement of a depression can be observed and recorded. Meteorologists are able to predict the kind of weather which will occur at places in the path of a moving depression. Depressions are sometimes called *cyclones*, but they should not be confused with the severe tropical storms which have the same name.

A depression's approach is indicated by a fall in the barometer because of the lowering of the air pressure. The sky becomes overcast and the atmosphere becomes warm and muggy. There is usually rain and drizzle.

An *anticyclone* is a moving region of high pressure. High-pressure air is more widely distributed. There are gentle breezes or calm weather. Slow-moving anticyclones bring long spells of cold, calm, frosty weather in winter and heat-waves in summer. There is little rainfall.

Rainfall

Rain is formed in three ways. Ascending currents of warm air rise and are cooled. Water vapour condenses and falls as rain. This is called *convectional* rain. *Relief* rain occurs when air is cooled in rising over high mountains. The rain falls on the *windward* side. *Cyclonic* rain occurs when warm air currents are forced over and cooled by masses of cold air.

Hailstones are frozen raindrops. *Snowflakes* are collections of tiny ice crystals formed by the cooling of water vapour in the atmosphere. They are always *symmetrical*. When you squeeze snowflakes together to make a snowball, they melt a little under the pressure of your hands. The water formed re-freezes, and so the snowball is held together.

Regelation

Show that ice melts under pressure. Attach equal weights to the ends of a thin copper wire. Hang the wire over a large block of ice. The wire slowly sinks as the ice below it melts under pressure. The water re-freezes, and so the block of ice remains in one piece. This phenomenon is known as *regelation*.

Ice and frost

Frost is dew and condensed water vapour which have frozen. *Hoar frost* is the white frost formed when water vapour in air below 0°C changes directly into a solid. *Icicles* are long spikes of ice that are formed when melting snow and ice re-freeze.

Thaws have a cooling effect. When snow and ice melt, they take heat (latent heat) from their surroundings.

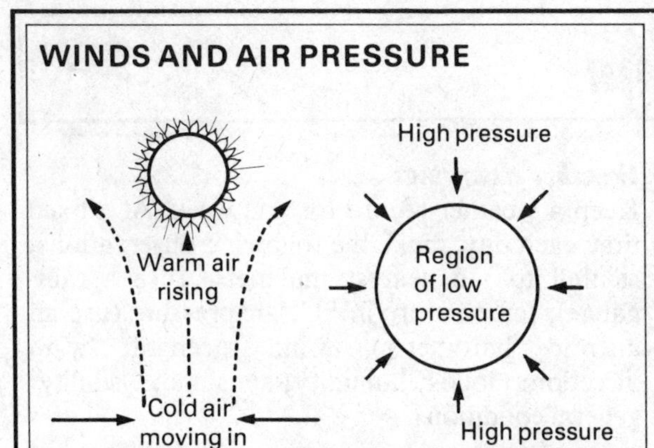
WINDS AND AIR PRESSURE
Warm air rising
Cold air moving in
High pressure
Region of low pressure
High pressure

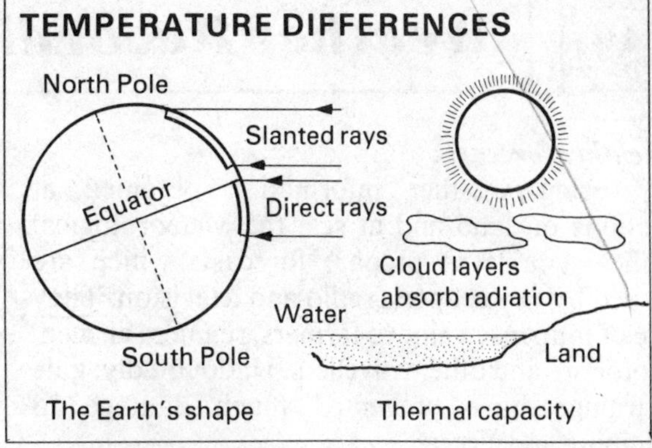
TEMPERATURE DIFFERENCES
North Pole
Slanted rays
Equator
Direct rays
South Pole
The Earth's shape
Cloud layers absorb radiation
Water
Land
Thermal capacity

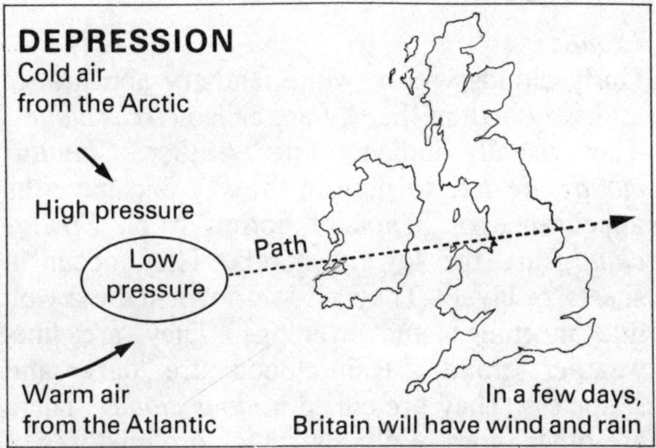
DEPRESSION
Cold air from the Arctic
High pressure
Low pressure
Path
Warm air from the Atlantic
In a few days, Britain will have wind and rain

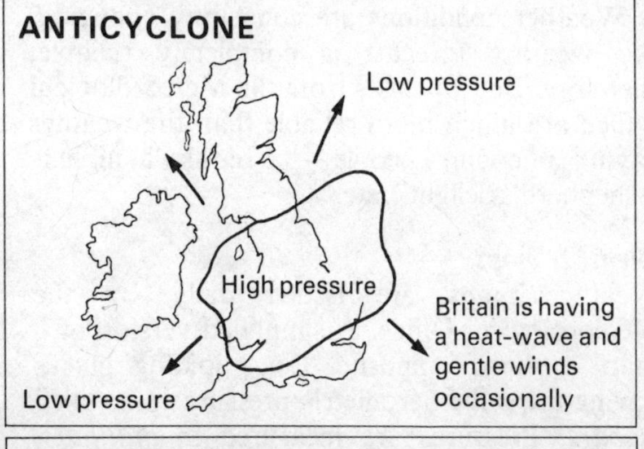
ANTICYCLONE
Low pressure
High pressure
Britain is having a heat-wave and gentle winds occasionally
Low pressure

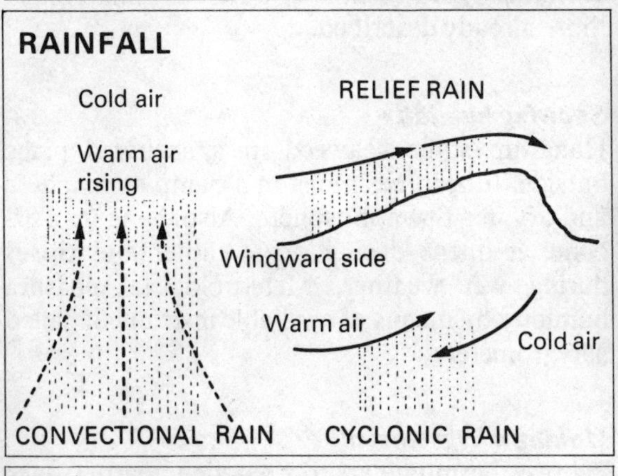
RAINFALL
Cold air
Warm air rising
CONVECTIONAL RAIN
RELIEF RAIN
Windward side
Warm air
Cold air
CYCLONIC RAIN

SNOWFLAKES
Notice the symmetrical shapes

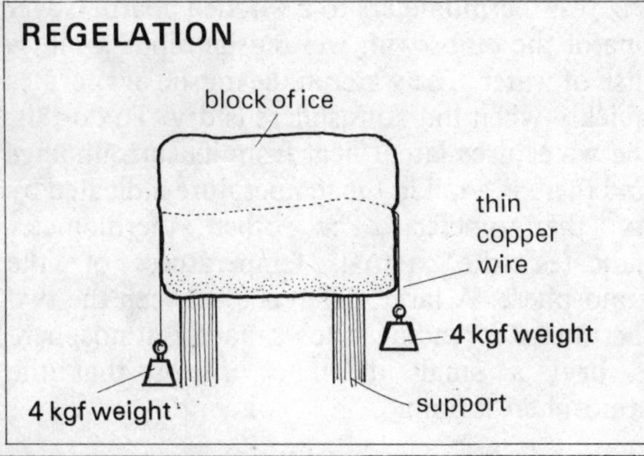
REGELATION
block of ice
thin copper wire
4 kgf weight
4 kgf weight
support

THAWS
Cold air
Melting snow and ice take heat from their surroundings

16 Weather Recording

Weather forecasts

By using weather information obtained at stations on land and at sea, the Meteorological Office is able to prepare forecasts which are issued by newspapers, radio and television. They are of immense value to farmers, seamen, airmen, motorists and other travellers. Undoubtedly, gale warnings have prevented much damage to shipping and aircraft.

Weather conditions are constantly changing. No weather forecast is completely reliable, therefore. But forecasts from the Meteorological Office are much more reliable than the weather sayings of country people—"A red sky at night is a shepherd's delight", etc.

Weather maps

Weather maps are issued daily by the Meteorological Office. A simplified version of a map is shown opposite. Lines joining places having the same barometric pressures are called *isobars*. Pressures are measured in *millibars*. (1000 millibars = 100 000 Pa = 75 cm of mercury.) Temperatures are given in °C. The directions and strengths of winds are shown by symbols with tails. Circles containing symbols indicate the positions of weather stations and the weather they are having—thunder, hail, etc.

Weather instruments

The instruments used in weather stations include rain-gauges, maximum and minimum thermometers, wet and dry bulb *hygrometers, sunshine recorders, anemometers* and barometers.

A maximum and minimum thermometer indicates the highest and the lowest temperatures reached during a day. A hygrometer indicates the humidity of the atmosphere. The hours of sunshine are shown by a scorch mark on a paper strip beneath the glass ball of a sunshine recorder. An anemometer indicates the wind direction and speed.

Weather-vanes indicate the wind direction. *Wind-socks* are used on some airfields to indicate the wind strength and direction. They are large, and so they are easily seen by pilots.

Weather recording

Keep a weather record for one week. At a fixed time each day, make the following observations: rainfall to the nearest millimetre (use a rain-gauge); temperature in °C; air pressure (use an aneroid barometer); wind strength; wind direction; clouds; humidity; sunshine; visibility; general conditions.

Clouds

Curly clouds with a white feathery appearance and very high in the sky are called *cirrus clouds*. They usually indicate fine weather. *Cumulus clouds* are not so high in the sky and have the appearance of heaps of cotton wool. *Stratus clouds* are the lowest clouds. They occur in sheets, or layers. They are low down in the sky on fine mornings and evenings. They are fine-weather clouds. Rain-clouds are dark and shapeless. They are called *nimbus clouds*. There are other types of clouds, which are mixtures of those already described.

Showing humidity

Hang up some seaweed in a sheltered place outside. It becomes moist in a damp atmosphere and dry in a dry atmosphere. Also, hang up a fir-cone. It opens during dry weather and closes during wet weather. Meteorologists measure humidity by means of a reliable instrument called a hygrometer.

Making a hygrometer

Fix two thermometers to a wooden board. Cover one of the bulbs with wet muslin dipping into a dish of water. The water in the muslin evaporates quickly when the atmosphere is dry. To do this, the water uses latent heat from its surroundings and there is a fall in the temperature indicated by its thermometer. The other thermometer indicates the actual temperature of the atmosphere. A large difference between the two thermometer readings shows that the atmosphere is dry; a small difference shows that the atmosphere is humid.

WEATHER MAP

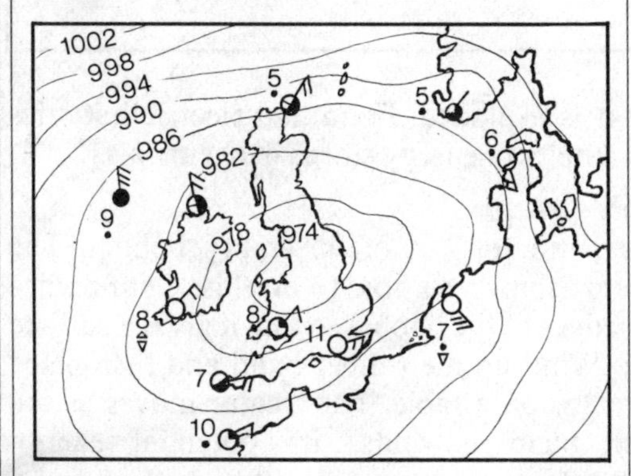

- ○ Cloudless sky
- ◑ Sky, half clouded
- ◕ Sky, three quarters clouded
- ⊗ Sky obscured
- ✳ Snow
- = Mist
- ≡ Fog
- • Rain
- ☈ Thunderstorm
- Rain shower
- ◎ Calm
- Hail shower
- Rain and snow

Wind speed symbols

1-2 knots 3-7 knots

8-12 knots 13-17 knots

Tail indicates wind direction

Each short tail represents 5 knots

Above 48 knots

WEATHER INSTRUMENTS

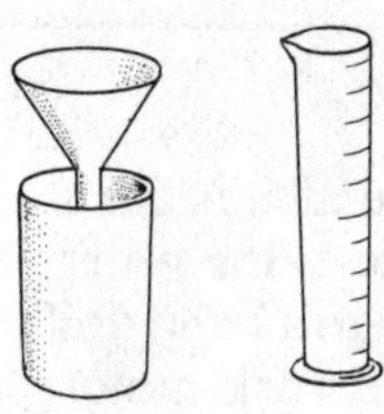
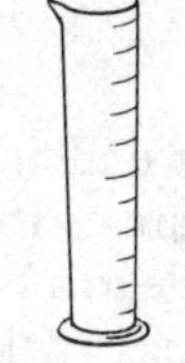

Rain-gauge

Maximum and minimum thermometer

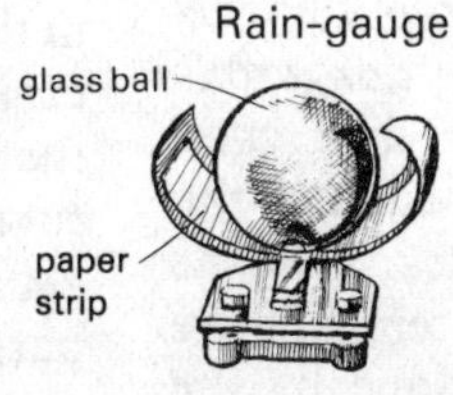

Sunshine recorder

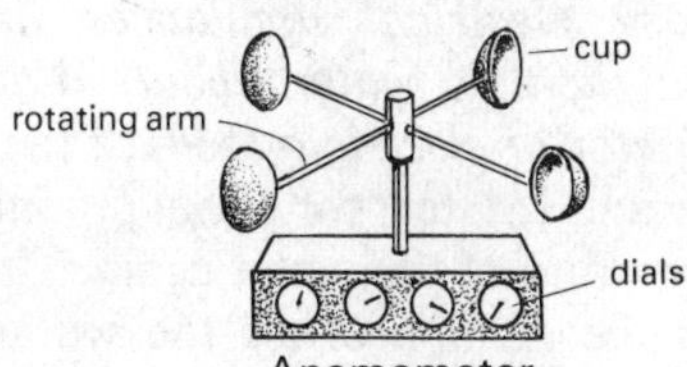

Anemometer

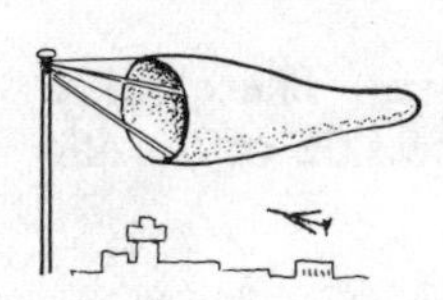

Wind-sock

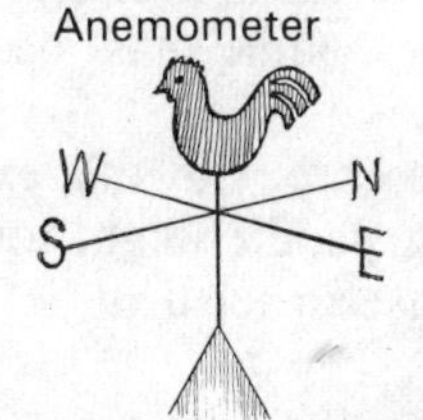

Weather-vane

SHOWING HUMIDITY

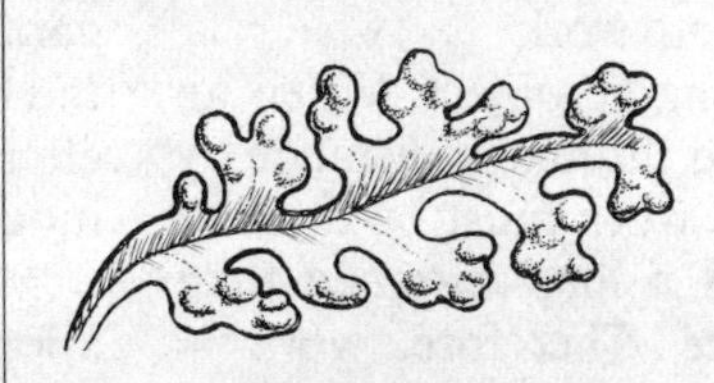

Seaweed

Fir-cone

WEATHER RECORDING

WEEKLY CHART							
Date							
Time							
Rainfall, mm							
Temperature, °C							
Air pressure, mm							
Wind direction							
Wind strength							
Clouds, types							
General weather							
Humidity							
Sunshine							
Visibility							
Remarks							

Wind direction: NE, SW, etc.
Wind strength: calm, light, fresh, strong, etc.
General weather: clear sky, blue sky, ràin, drizzle, snow, overcast sky, partly cloudy, hail, fog, mist, thunder, etc.
Humidity: wet, fair, dry, etc.
Sunshine: dull, fair, bright, very bright, etc.
Visibility: good, moderate, bad, etc.

CLOUD TYPES

Cirrus

Cumulus

Stratus

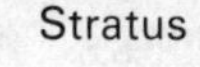

Nimbus

MAKING A HYGROMETER

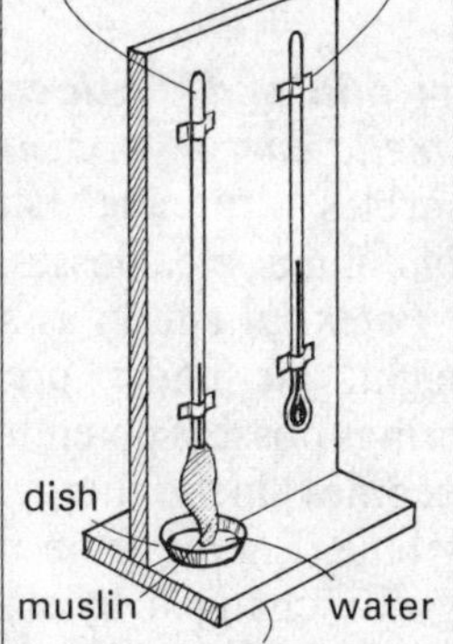

17 Energy and Work

Heat energy and work

Pour water into a beaker or a jar until it is about 10 cm deep. Take the temperature of the water. Attach a wooden or plastic paddle to a hand drill or the spindle of a powerful clockwork motor. *(An electric drill should not be used for this because electrical appliances are dangerous where there is water about. Water is a good conductor of electricity!)* Place the paddle in the water and rotate it for about 5 minutes. Take the temperature of the water again. The temperature will have risen because the work done by the moving paddle has been changed into heat energy.

Take care with this experiment. The quickly rotating paddle might touch the side of the jar and cause an accident.

Energy and work

When *work* is done, *energy* is used. In a diesel engine, for example, burning oil supplies heat energy which is changed into work.

Work is done in moving an object. When an object is lifted, it is moved through a distance by an *effort,* or *force.* More work must be done to raise the object through a long distance than through a short distance. Therefore, work = effort × distance through which the effort moves.

Force is measured in *newtons* (N). 10 N ≙ 1 kgf. Work is measured in *newton metres* (N m) or *joules* (J). Thus, the work done in lifting a 4 kgf weight for a distance of 5 m = 20 kgf m = 200 N m = 200 J.

Potential and kinetic energy

Kinetic energy is energy of movement. Rolling marbles, running streams, swinging clock pendulums, etc., possess kinetic energy.

Potential energy is stored energy. Water at a height, air under pressure and coiled watch springs possess potential energy. Potential energy becomes kinetic energy when falling water turns a turbine. Energy from the Sun, stored in oil ages ago, is changed by a diesel engine into kinetic energy. Gunpowder produces kinetic energy when it is exploded. Food and electric batteries are examples of energy stored in chemicals.

A bobbin tractor

Make a tractor out of an empty bobbin. The diagrams show you how to do this. Cut notches in the edges of the bobbin so that it grips a surface easily. Wind up the rubber band and then place the tractor on a table. The tractor moves as the rubber band unwinds. Its potential energy becomes kinetic energy. What do you think is the purpose of the piece of candle shown in the diagram?

Inertia

More effort is required to move a heavy object than a light object. More effort is required to bring a heavy moving object to a standstill than a light object moving at the same speed. The resistance offered by an object to a force applied to it is called its *inertia.*

A person standing in a bus moves forward quickly when the bus stops suddenly. His inertia carries him forward. He moves backwards when the bus starts suddenly. A locomotive uses more effort to start a train than it does to keep it in motion.

Moving objects come to rest, even when no apparent force is applied, because of the force exerted by gravity and *friction.*

Showing inertia

Place a coin on a card covering a beaker. Flick the card away quickly. The coin drops into the beaker. The coin does not move forward with the card because of its inertia.

Place a coin on a flat piece of wood. Quickly slide the wood forward against an obstacle. The wood stops but the coin moves forward because of its inertia.

Spin two eggs, one raw and the other hard-boiled, on a table. Place your fingers on the eggs for a moment. When you remove your fingers, the hard-boiled egg remains still but the raw egg revolves again slowly. The liquid inside the raw egg has continued to revolve because of its inertia, and so, when your finger is removed, the shell is pulled around with it.

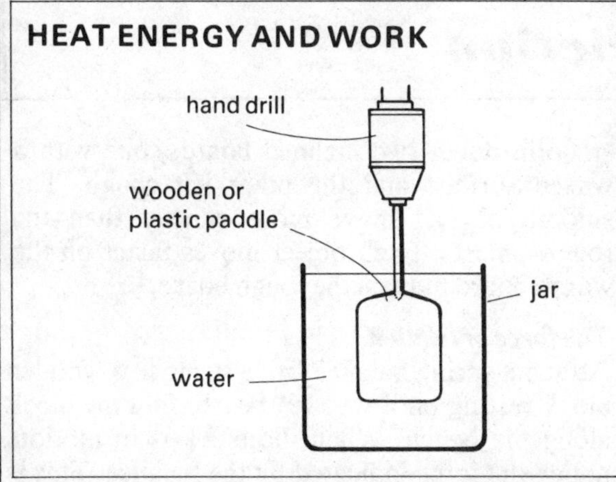
HEAT ENERGY AND WORK
hand drill
wooden or
plastic paddle
jar
water

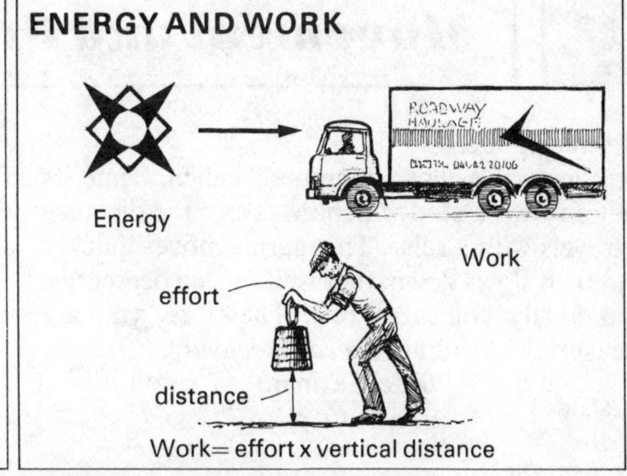
ENERGY AND WORK
ROADWAY
HAULAGE
Energy
Work
effort
distance
Work= effort x vertical distance

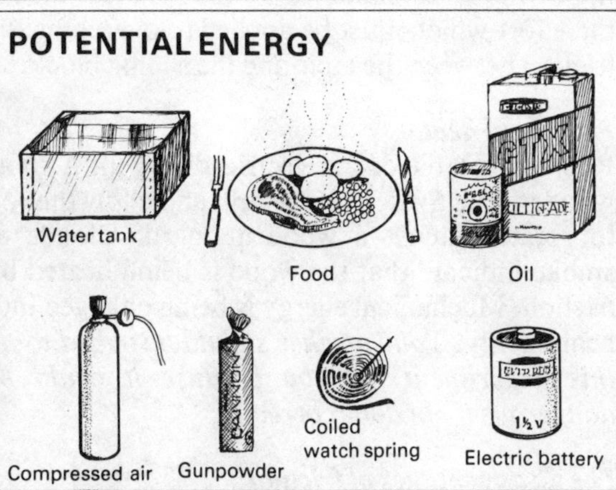
POTENTIAL ENERGY
Water tank
Food
Oil
Compressed air
Gunpowder
Coiled
watch spring
1½ v
Electric battery

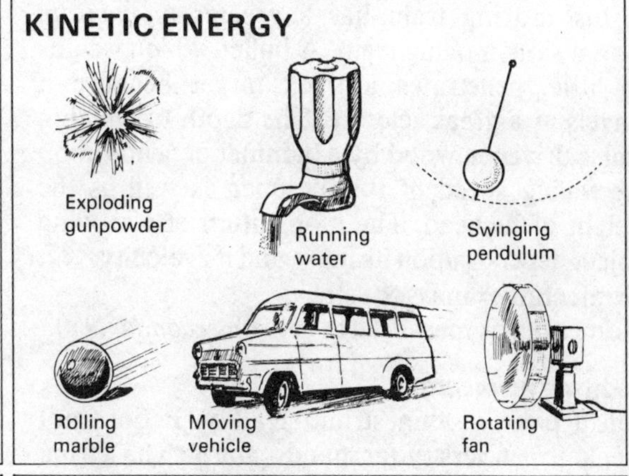
KINETIC ENERGY
Exploding
gunpowder
Running
water
Swinging
pendulum
Rolling
marble
Moving
vehicle
Rotating
fan

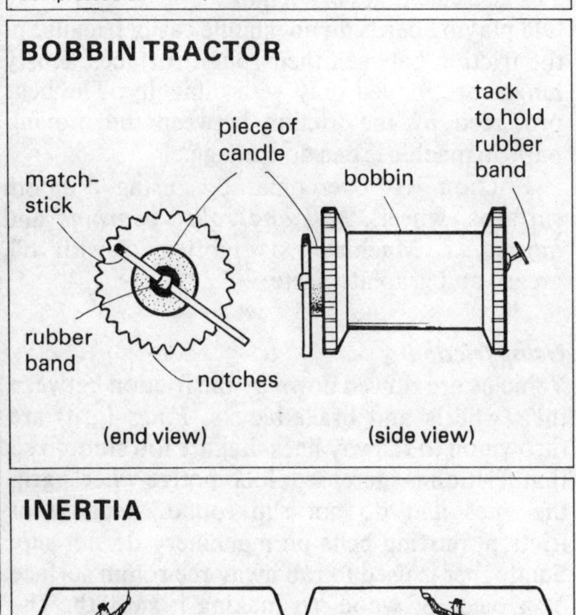
BOBBIN TRACTOR
piece of
candle
tack
to hold
rubber
band
bobbin
match-
stick
rubber
band
notches
(end view)
(side view)
INERTIA
Bus stopping
Bus starting

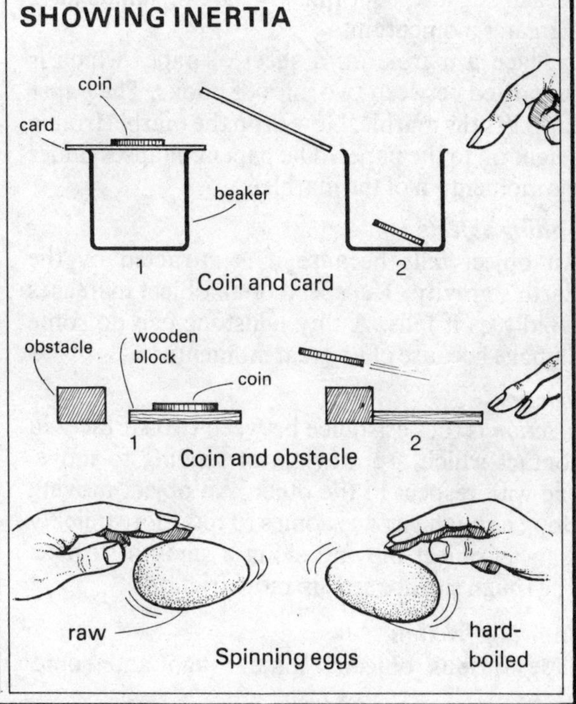
SHOWING INERTIA
coin
card
beaker
1
2
Coin and card
obstacle
wooden
block
coin
1
2
Coin and obstacle
raw
Spinning eggs
hard-
boiled

18 Movement and Friction

Measuring speed
Roll a marble along a smooth bench. Time its motion with a stop-watch. Measure the distance it travels with a ruler. The marble moves quickly and then slows down, that is, its speed decreases, and finally comes to rest. Therefore, you are measuring its average *speed*, or *velocity*.

$$\text{Average speed} = \frac{\text{distance (in m)}}{\text{time (in s)}}\ \text{m/s}$$

Momentum
A fast-moving train has a greater *momentum* than a slow-moving train. A bullet, which weighs so little, penetrates a thick target because it travels at a great velocity. The depth to which a nail is driven in wood by a hammer depends upon the falling speed of the hammer as well as the weight of its head. The momentum of a moving object depends upon its mass and its velocity.

Momentum = mass × velocity

Momentum is measured in *newton seconds* (N s).

Showing momentum
Pull a piece of thin string steadily. It does not break. Now jerk your hands apart. The string breaks because your quickly moving hands have a greater momentum.

Place a marble on a sheet of paper which is supported between two piles of books. The paper supports the marble. Now drop the marble from a height on to the paper. The paper collapses under the momentum of the marble.

Falling objects
An object falls because it is attracted by the Earth's gravity. The speed of an object increases rapidly as it falls. A tiny hailstone can do some damage because of its great momentum.

Friction
Friction is the resistance between two surfaces in contact which are moving, or tending to move, one with respect to the other. An object moving along a rough surface comes to rest more quickly than an object moving along a smooth surface. The rough surface sets up more friction.

Showing friction
Slide various objects, some rough and some smooth, down two inclined boards, one with a waxed surface and the other left rough. The smooth objects move more quickly than the rough objects. Each object moves faster on the waxed board than on the rough board.

The force of friction
Attach a spring balance and a string to a wooden block resting on a wooden bench. Pull the block along the bench. When the block is in motion, notice the force indicated by the balance. This is the effort which must be applied to overcome the friction between the table and the sliding block.

Frictional heat
Replace the bit of an electric drill with a short wooden rod. Switch on the drill and allow the rod to rotate against a wooden board. Clouds of smoke indicate that the wood is being heated by friction. Mechanical energy is being changed into heat energy. *Your teacher should be asked to do this experiment for you because it could be dangerous if not done correctly.*

The disadvantages of friction
Old playing cards do not shuffle easily because of the friction between their rough surfaces. Rusty hinges are moved only with difficulty. The heat produced by the friction between the moving parts in machines can do damage.

Friction is overcome by using smooth surfaces, wheels, *ball and roller bearings* and *lubricants*. Machinery is lubricated with oil, grease and graphite paste.

Using friction
Vehicles are slowed down by the friction between their wheels and brake-blocks. *Fines* (grit) are thrown on to railway lines that are too smooth so that friction is increased; locomotive wheels grip the lines and do not slip round. Because of friction, moving belts on machinery do not slip. Sandpaper is used to rub away the rough surface of a piece of wood, so making it smooth. The treads on motor-vehicle tyres enable them to grip the surface of a road; smooth tyres could cause the wheels to slip.

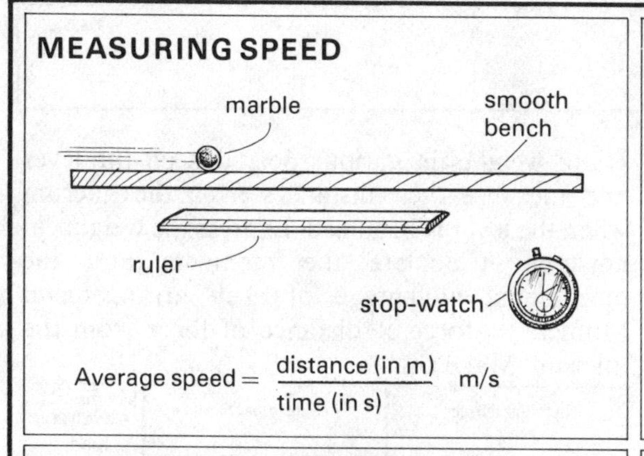
MEASURING SPEED
marble
smooth bench
ruler
stop-watch
Average speed = distance (in m) / time (in s) m/s

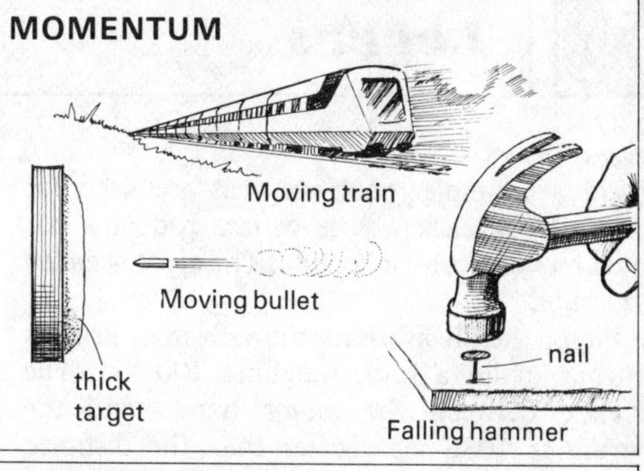
MOMENTUM
Moving train
Moving bullet
thick target
nail
Falling hammer

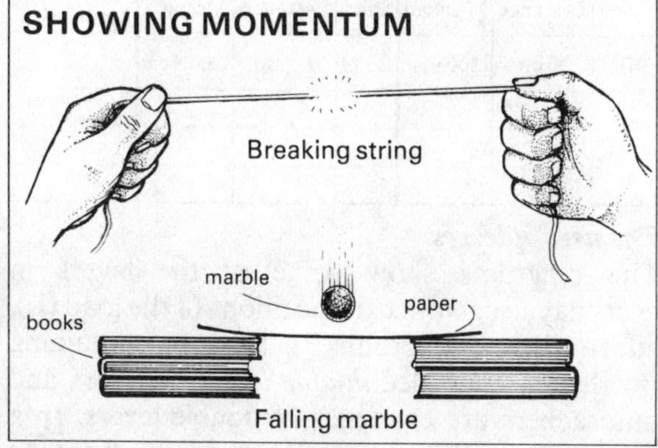
SHOWING MOMENTUM
Breaking string
marble
paper
books
Falling marble

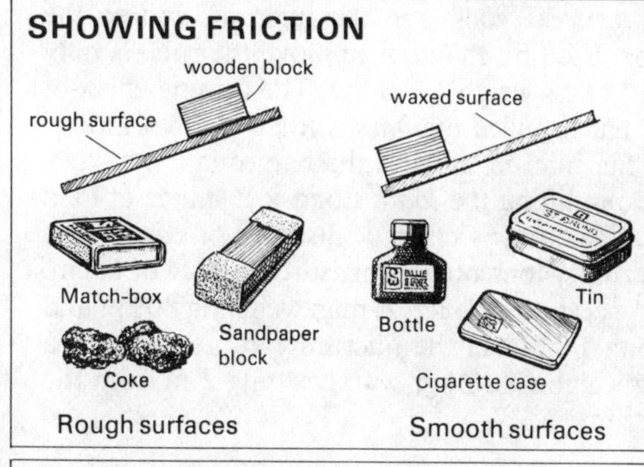
SHOWING FRICTION
wooden block
rough surface
waxed surface
Match-box
Sandpaper block
Coke
Bottle
Tin
Cigarette case
Rough surfaces
Smooth surfaces

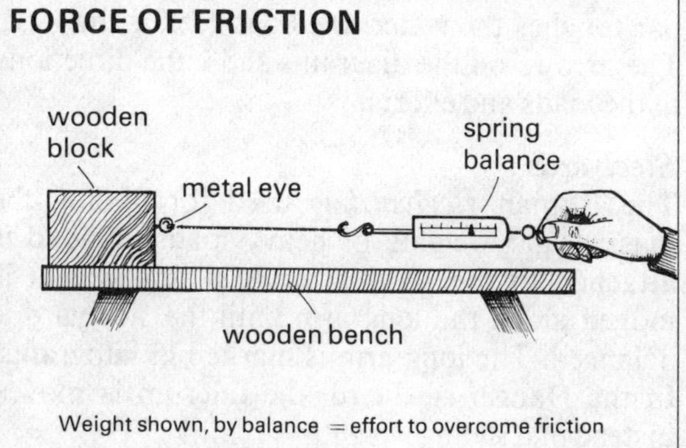
FORCE OF FRICTION
wooden block
metal eye
spring balance
wooden bench
Weight shown, by balance = effort to overcome friction

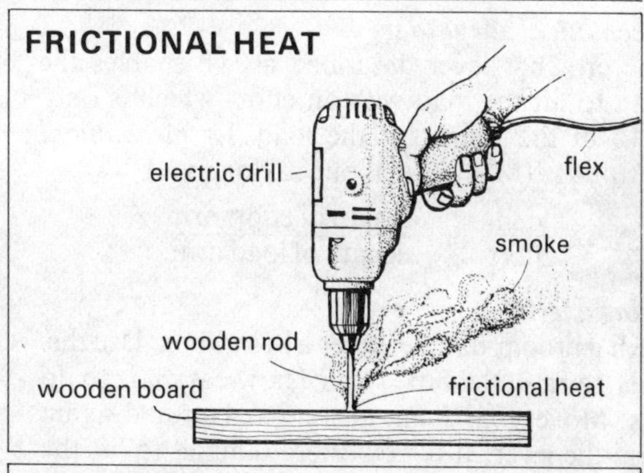
FRICTIONAL HEAT
electric drill
flex
smoke
wooden rod
wooden board
frictional heat

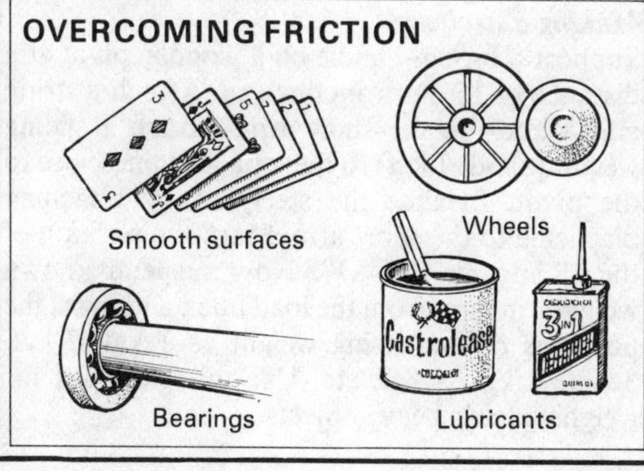
OVERCOMING FRICTION
Smooth surfaces
Wheels
Castrolease
3 in 1
Bearings
Lubricants

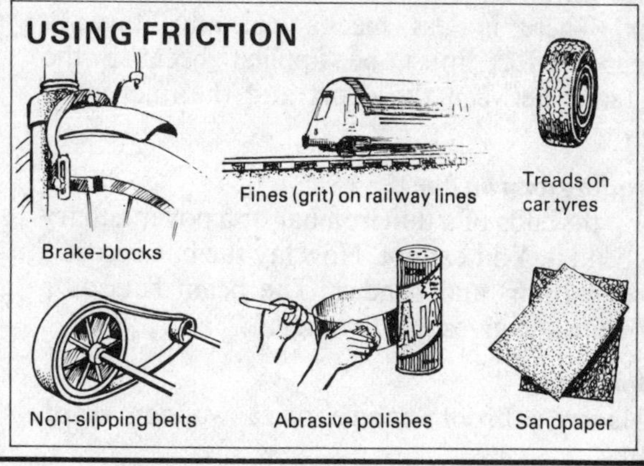
USING FRICTION
Fines (grit) on railway lines
Treads on car tyres
Brake-blocks
AJAX
Non-slipping belts
Abrasive polishes
Sandpaper

19 Levers

Levers

Levers are simple machines that are used for making work easier. A lever is a rod or a bar which moves freely on a pivot. The pivot is called a *fulcrum.*

One of the illustrations shows a man using a crowbar to lift a rock weighing 100 kgf. The distance between the man's hands and the fulcrum is 10 times greater than the distance between the rock and the fulcrum, and so the effort used by the man to move the rock is only 1/10 of its weight—10 kgf. The turning effect of the bar is called the *moment*, and the work done by the man in moving the bar equals the work done in lifting the *load.* Load × distance of load from fulcrum = effort × distance of effort from fulcrum. Moments are measured in N m or N cm.

A seesaw is a lever. A man weighing 80 kgf and sitting 1 m from the fulcrum can just balance a boy weighing 40 kgf who is sitting 2 m from the fulcrum.

Mechanical advantage

The crowbar lever described above enables the man to lift the load with an effort which is only 1/10 of the weight of the load. Its *mechanical advantage* (M.A.), or labour-saving, is 10.

$$\text{M.A.} = \frac{\text{load}}{\text{effort}} \text{ or } \frac{\text{length of effort arm}}{\text{length of load arm}}$$

Using a lever

Push a broom-handle under a heavy box. Use this lever to raise the box. Notice how easy it is to do this. Move your hand nearer the box and again raise the lever. It is now more difficult to lift the box. There is less mechanical advantage; a greater effort must be applied because the distance between the effort and the fulcrum is less.

Bending an iron bar

Grip the ends of a thin iron bar or a poker and try to bend it. You cannot. Now lay the bar across a strong beam and bend it. The beam acts as a fulcrum and gives you *leverage.*

Moments

Balance a broom-handle on a wooden pivot. Hang weights in various positions on the lever and measure their distances from the fulcrum when the lever is balanced. Express the weights in newtons. Calculate the moments and the mechanical advantage of each arrangement. Moment = force × distance of force from the fulcrum. Make a table.

Left-hand Side			Right-hand Side			Mechanical Advantage
Load	Distance	Moment	Effort	Distance	Moment	load ÷ effort
40N	30 cm	1200N cm	20 N	60 cm	1200N cm	2

The uses of levers

The diagrams show some of the levers in everyday use. Notice the positions of the load (L), effort (E) and fulcrum (F). Devices containing two levers are called *double-levers*. Scissors and nutcrackers are examples of double-levers. In a rowing boat, the fulcrum is at the place where the oar touches the water; the boat itself is the load. The arrows on the diagrams show the directions of the loads and efforts.

Steelyards

The Roman *steelyard* is used for finding the masses and weights of heavy loads. A load is attached to the short arm. A sliding weight is moved along the long arm until the steelyard is balanced. The long arm is marked in kilograms. In the Danish steelyard, the fulcrum is moved instead of a sliding weight.

Making a steelyard

Support a broom-handle on a wooden pivot at a distance of 15 cm from one end. Attach a string and a hook to the short arm. Attach a sliding weight (about 1 kgf) to the long arm and close to the pivot. Balance the steelyard by attaching plasticine to the short arm. Mark the position of the sliding weight "O". Now suspend known weights, in turn, from the load hook and mark the positions of the sliding weight as 1 kgf, 2 kgf, 3 kgf, 4 kgf, 5 kgf, etc. Use this steelyard for weighing fairly heavy objects.

LEVERS

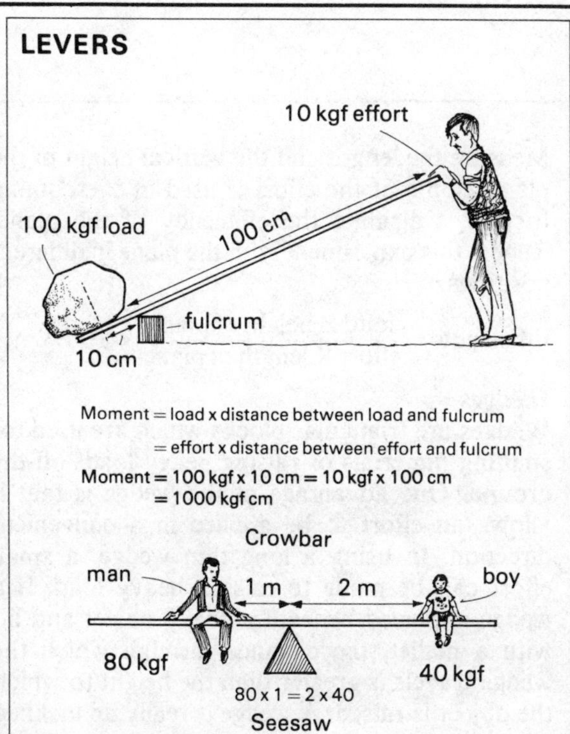

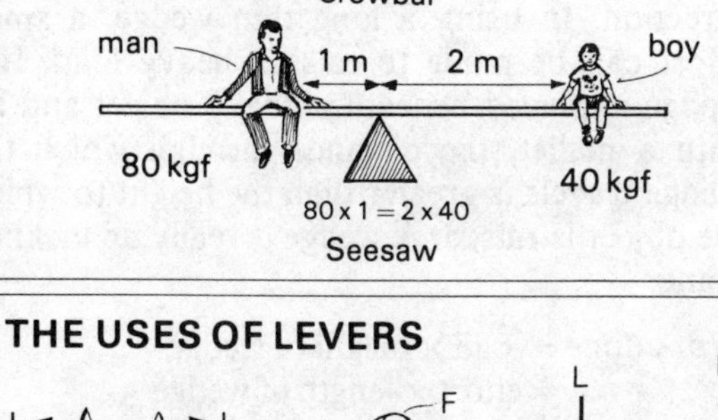

USING A LEVER

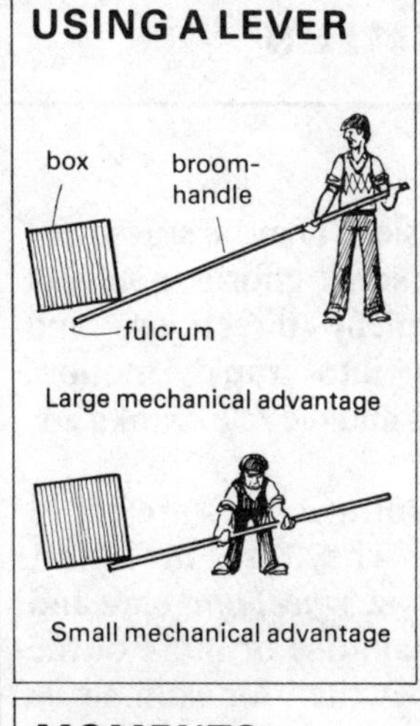

BENDING AN IRON BAR

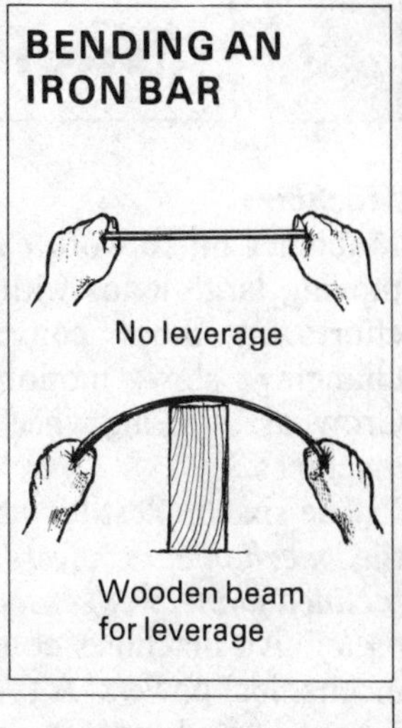

MOMENTS

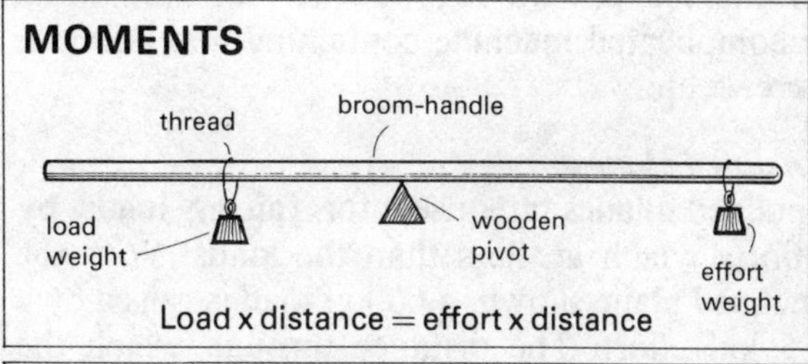

THE USES OF LEVERS

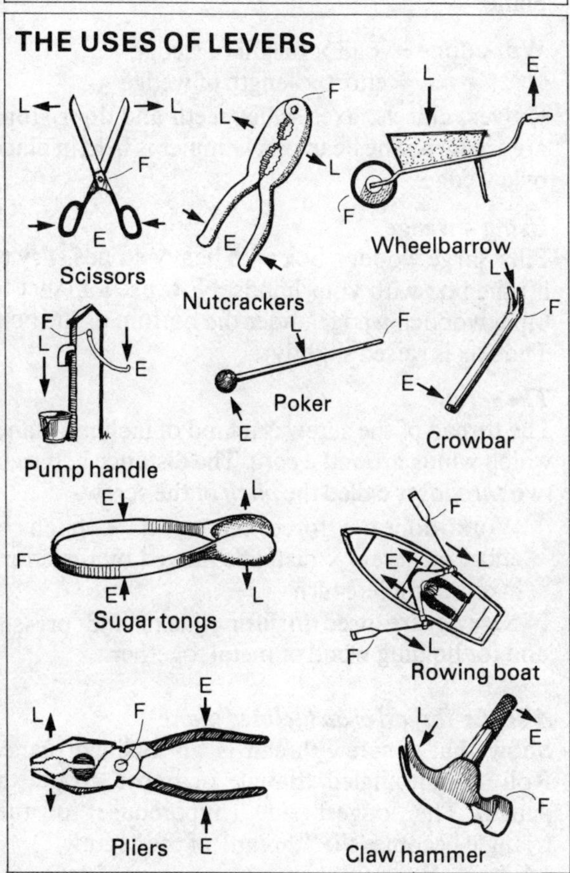

STEELYARDS

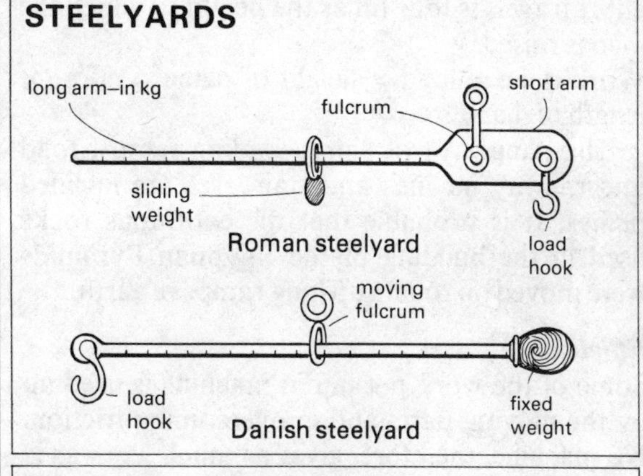

MAKING A STEELYARD

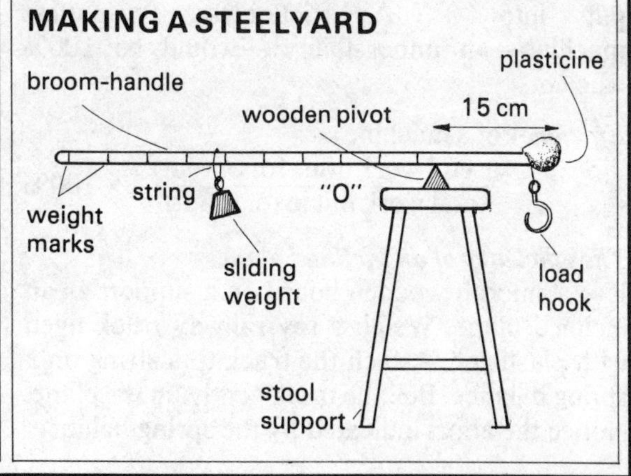

20 Machines

Machines

Machines make work easier. They are used for moving large loads with small efforts, applying efforts in more convenient directions, and changing slow motion into rapid motion. Crowbars, steering-wheels and bicycle cranks are machines.

The six simplest forms of machines are called the *mechanical powers*. They are the *lever, inclined plane, wedge, screw, wheel and axle* and *pulley*. All machines contain one or more of the mechanical powers. A typewriter, for example, is a complicated machine containing many levers, screws, etc.

Inclined planes

Inclined planes are used for raising loads by efforts which are less than the loads. With the inclined plane shown, a 60 kgf load is raised by a 15 kgf effort. The distance through which the effort travels is four times the height to which the load is raised.

Work done = load × height of plane = effort × length of plane

The gangways of ships, loading ramps, road and railway inclines and staircases are inclined planes. It is probable that the enormous rocks used in the building of the Egyptian Pyramids were moved on rollers up long ramps of earth.

Efficiency

Some of the work put into a machine is used up by the moving parts and in overcoming friction. No machine, therefore, gives as much work as is put into it. A frictionless, weightless machine—an impossibility!—would be 100% efficient.

Efficiency of a machine =

$$\frac{\text{useful work done (or output)}}{\text{total work put in (or input)}} \times 100\%$$

The efficiency of an inclined plane

Use a smooth wooden board on a support as an inclined plane. Weigh a toy railway truck filled with plasticine. Attach the truck to a string on a spring balance. Pull the truck gently up the plane. Notice the effort indicated by the spring balance. Measure the length and the vertical height of the plane. Some of the effort is used in overcoming friction. Calculate the efficiency of the plane. Repeat this experiment with the plane in different positions.

$$\text{Efficiency} = \frac{\text{load} \times \text{height of plane}}{\text{effort} \times \text{length of plane}} \times 100\%$$

Wedges

Wedges are triangular blocks which are used for splitting materials or raising heavy loads off the ground. One advantage of the wedge is that it allows an effort to be applied in a convenient direction. In using a long thin wedge, a small effort can be made to raise a heavy load. If a wedge is placed beneath a heavy object and hit with a mallet, the distance through which the wedge travels is greater than the height to which the object is raised. A wedge is really an inclined plane.

Work done = load × height of wedge
= effort × length of wedge

Knives, chisels, axes, nails, teeth and door-stops are wedges. The head of a hammer is held in place by a wedge.

Using a wedge

Fill a large wooden box with heavy stones. Try to lift the box with your hands. Now use a mallet to tap a wooden wedge under the bottom of the box. The box is raised slightly.

The screw

The thread of the screw is a kind of inclined plane which winds around a *core*. The distance between two *threads* is called the *pitch* of the screw.

Work done = force on wood × pitch = effort on screw × distance moved by the effort in one complete turn

Screws are used in lifting-jacks and presses and for holding wood or metal together.

A screw-thread is an inclined plane

Show that a screw-thread is an inclined plane. Roll a right-angled triangle of paper around a pencil. The longest side (hypotenuse) of the triangle becomes the "thread" of the "screw".

MACHINES

INCLINED PLANE

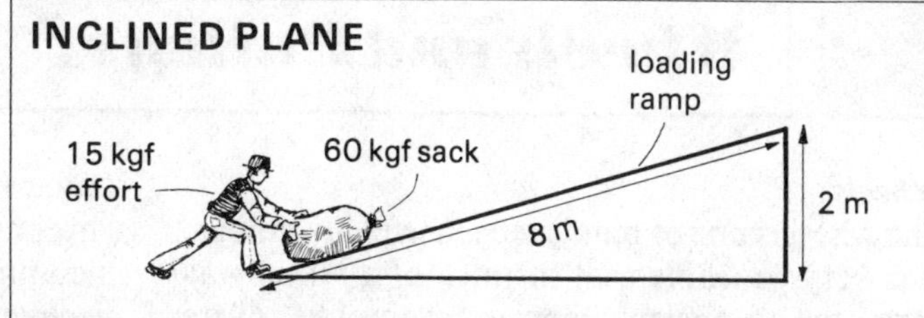

THE MECHANICAL POWERS

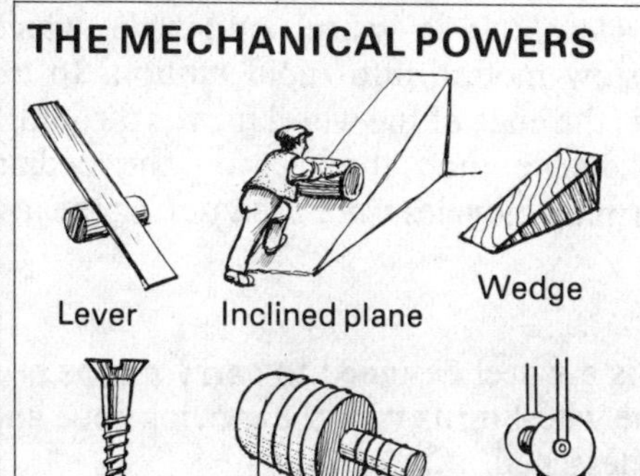

INCLINED PLANES

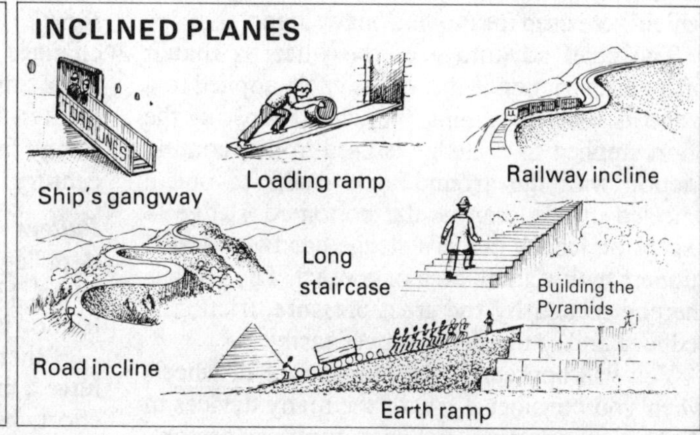

EFFICIENCY OF AN INCLINED PLANE

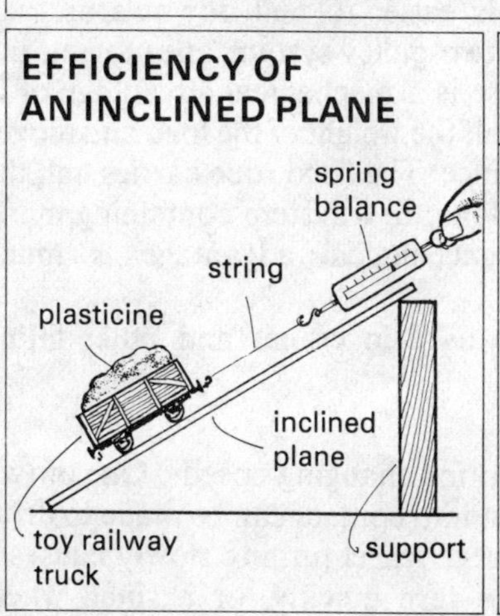

WEDGES

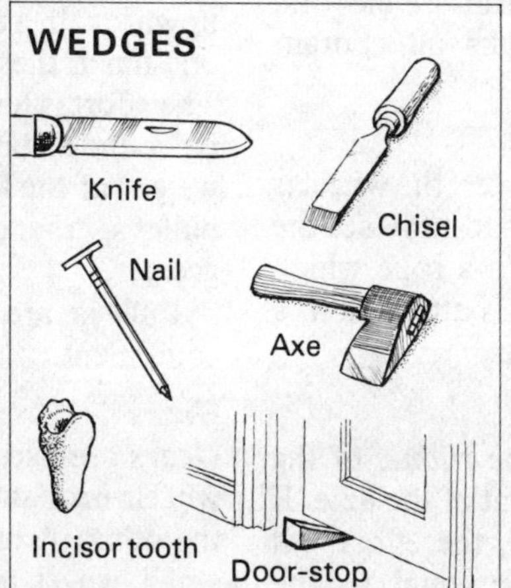

USING A WEDGE

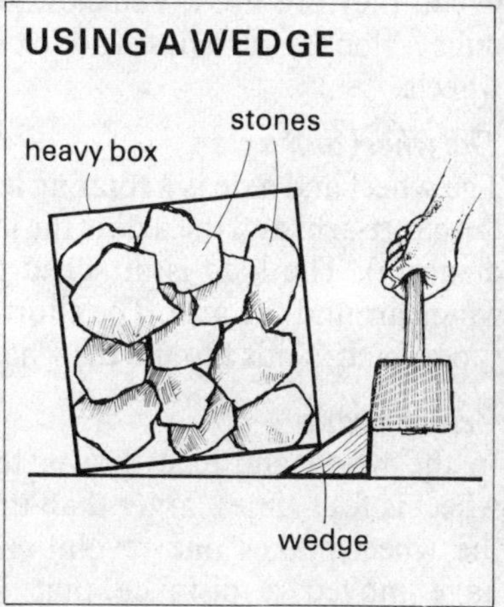

SCREW

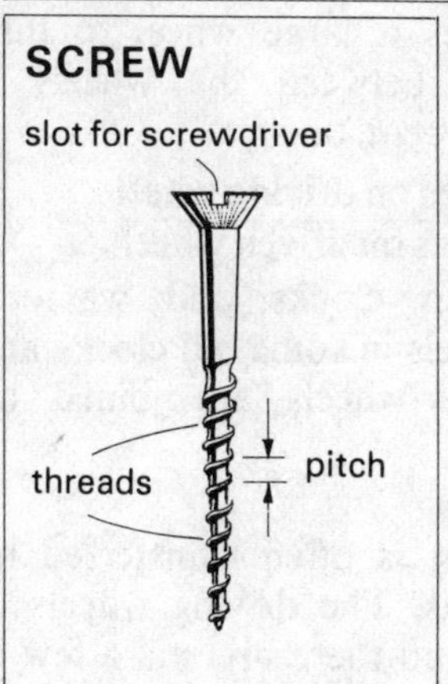

CAXTON'S PRINTING PRESS

Used by William Caxton, 1422-1491

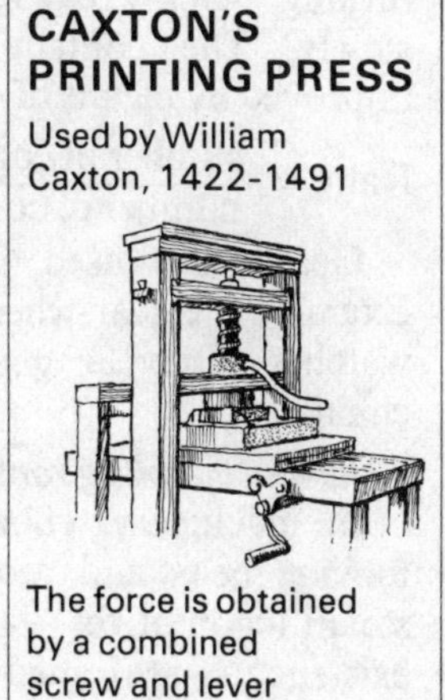

The force is obtained by a combined screw and lever

SCREWS ARE INCLINED PLANES

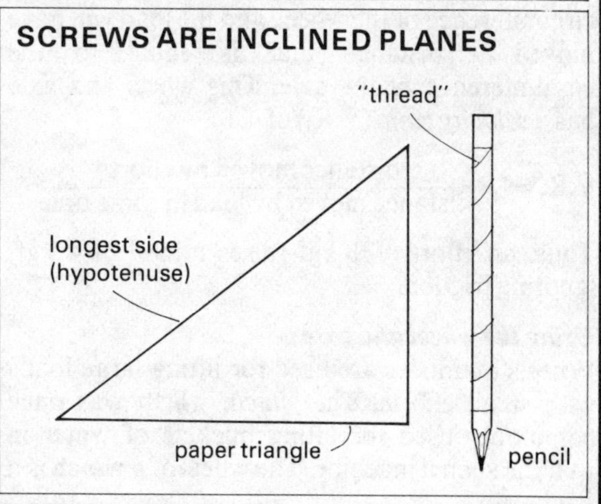

21 Wheels and Pulleys

Wheels

The wheel is one of man's most useful inventions. It is very probable that the idea of a wheel was suggested to ancient peoples by the log rollers which were used for moving heavy loads.

The chief advantage of the wheel is that it offers little friction. Most of the effort applied to a wheel is used in turning the wheel. Most of the effort applied to a sledge is used in overcoming friction with the ground. It is easier to pull a wheeled truck than a flat-bottomed sledge – except on ice! A modern sledge has two parallel runners with a small area of contact. The snow is thereby melted by the great pressure, friction is reduced and so the sledge moves easily.

You will appreciate the importance of wheels when you consider a few of the many devices in which they are used. Vehicles, furniture castors, pulley-blocks, turbines and clocks all contain wheels.

The wheel and axle

The wheel and axle is a rotating lever; the wheel is the effort-arm and the axle is the load-arm (see the diagram). The load is attached to a rope which winds around the axle. The effort is attached to a rope which winds around the wheel.

Velocity ratio

In the wheel and axle shown, the *radius* of the wheel is four times larger than that of the axle. If the wheel makes one revolution, the effort will have moved a distance that is equal to the circumference of the wheel, and the load will have moved a distance that is equal to the circumference of the axle. This wheel and axle has a *velocity ratio* (V.R.) of 4.

$$\text{V.R.} = \frac{\text{distance moved by effort}}{\text{distance moved by load in same time}}$$

Thus, an effort of 1 kgf raises a load of 4 kgf, ignoring friction.

Using the wheel and axle

Wheels and axles are used for lifting large loads with small efforts. The *winch,* which was once commonly used for lifting buckets of water in wells, is such a machine. The wheel of a winch is a long arm with a handle. The *capstan* is a form of wheel and axle having several effort arms. It was in common use in the days of sailing ships, before mechanical power became available. The *penny farthing bicycle* is a wheel and axle which changes slow motion into rapid motion. In the same time, the edge of the wheel moves through a greater distance than the feet on the pedals. Penny farthing bicycles were used during the last century.

Pulleys

A pulley is a wheel designed to carry a rope or a chain. The wheel turns with the moving rope and friction is lessened.

If there were no friction, a single pulley would have a mechanical advantage of 1, that is, load = effort, but it is easier to pull downwards than upwards. In a two-pulley system, one moving and one fixed, there is a mechanical advantage of 2. The effort is half the weight of the load and moves twice the distance. The fixed rope carries half the weight of the load. In a system containing many pulleys, the mechanical advantage is much greater.

Pulleys are used in cranes and other lifting equipment.

Gears

Gears are used for changing speeds. One of two wheels brought into contact can be made to drive the other. A large wheel turning slowly causes a small wheel to turn quickly, or a small wheel turning quickly causes a large wheel to turn slowly. The contact between the wheels is improved by means of *teeth,* or *cogs.*

$$\text{Ratio} = \frac{\text{number of cogs on driving wheel}}{\text{number of cogs on driven wheel}}$$

Gears are used in clocks and watches. Examine the gear-wheels in some old clocks and watches. Bicycle gear-wheels are joined by chains.

Belts and coupling rods

In machinery, effort is often transferred by moving belts and rods. The driving wheels of steam locomotives—and there are still a few in existence!—are joined by coupling rods.

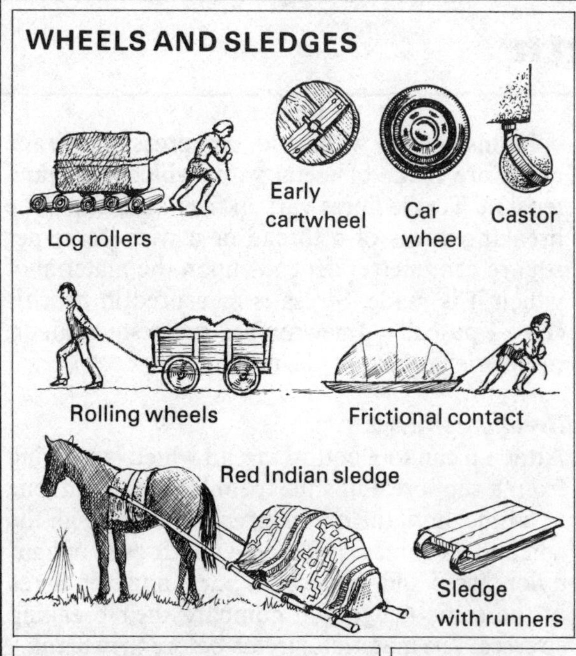
WHEELS AND SLEDGES
Log rollers
Early cartwheel
Car wheel
Castor
Rolling wheels
Frictional contact
Red Indian sledge
Sledge with runners

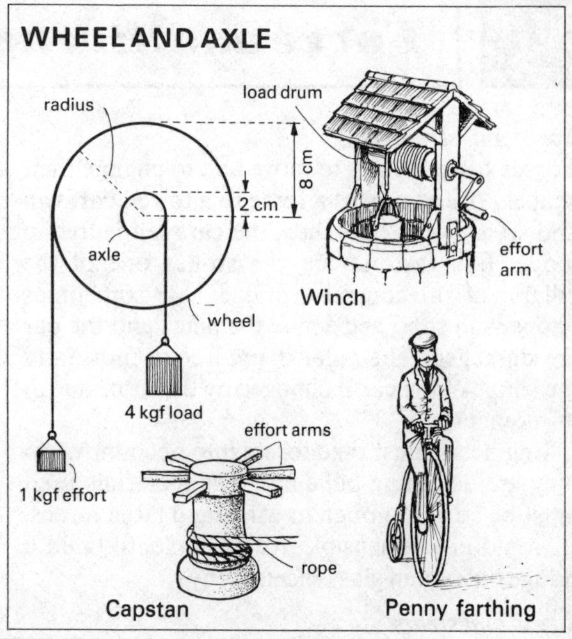
WHEEL AND AXLE
radius
load drum
8 cm
2 cm
axle
wheel
effort arm
Winch
4 kgf load
1 kgf effort
effort arms
rope
Capstan
Penny farthing

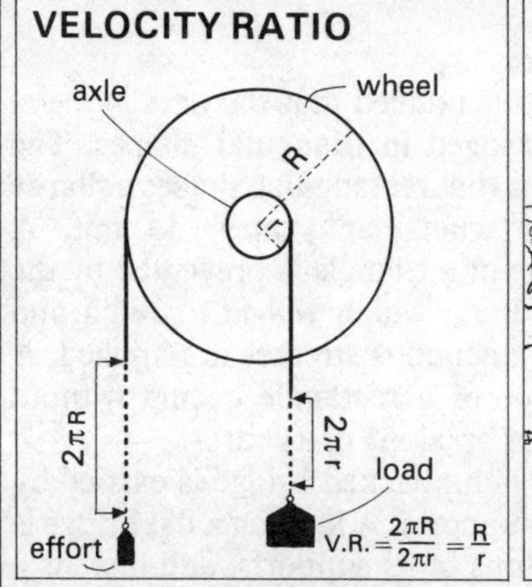
VELOCITY RATIO
axle
wheel
R
r
2πR
2πr
load
effort
V.R. = 2πR/2πr = R/r

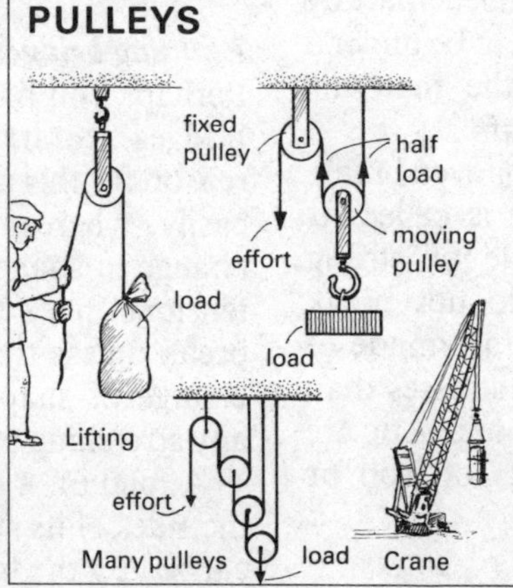
PULLEYS
fixed pulley
half load
effort
moving pulley
load
Lifting
load
effort
Many pulleys
load
Crane

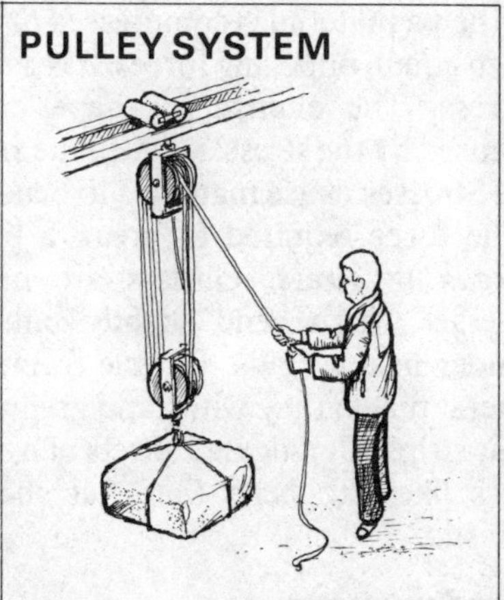
PULLEY SYSTEM

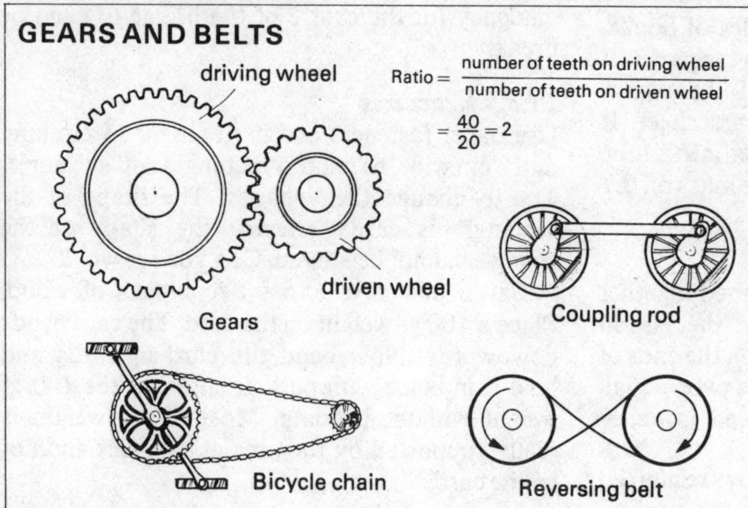
GEARS AND BELTS
driving wheel
Ratio = number of teeth on driving wheel / number of teeth on driven wheel
= 40/20 = 2
driven wheel
Gears
Coupling rod
Bicycle chain
Reversing belt

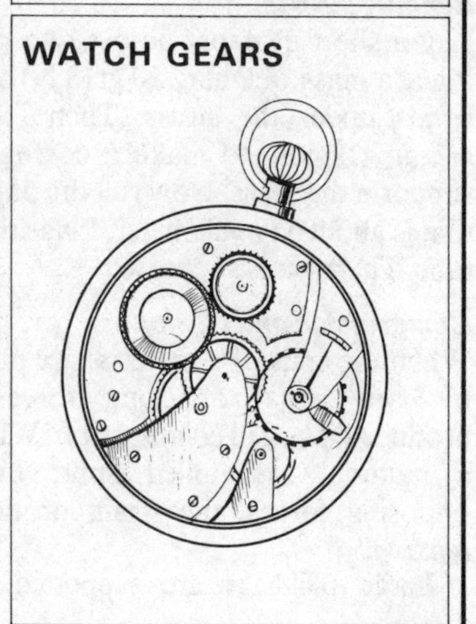
WATCH GEARS

22 Force and Strength

Force and strength

Forces cause bodies to move and to change their shapes. A car applies a force to a towed caravan and so it moves. But then, the caravan is free to move. If a fast-moving car strikes one of the pillars of a concrete bridge, as sometimes happens in road accidents, the pillar and the car are damaged. The pillar is not free to move and the shape of the car is changed by the force due to its momentum.

Engineers must take forces into account when they are designing buildings. The materials used must be strong enough to withstand large forces. It would not be sensible, for example, to build a bridge out of thin glass sheets. Why?

Stress and strain

When a material is compressed or pulled apart by two equal, opposing forces, it is said to be under a stress. The change of shape of the material, caused by the stress, is called the *strain*.

Stresses cause materials to bend or even break. The force required to break a bar is called its *breaking stress*. Girders are made of strong metals which bend slightly but do not break under heavy loads. Bicycle frames are made of metal tubes; they withstand greater stresses than flat strips. Corrugated sheets of iron or plastic are less likely to bend than flat sheets of iron or plastic.

Bending stresses

Lay a sheet of paper across two piles of books. Place a mass weighing 30 gf (3 N) on the paper. It bends under the stress. Then fold the paper, zigzag fashion, to make a corrugated sheet. It supports the mass. Now roll the paper into a tube that is about 3 cm diameter. Place the mass on the tube. The mass is supported.

Compression and tension

When the ends of a material are pushed together by two equal opposing forces, the strain produced is called *compression*. When the ends of a material are pulled apart by two equal, opposing forces, the strain produced is called *tension*.

Large buildings are supported by reinforced columns which withstand compression. Crane ropes are made of metal wires which withstand tension. Textile fibres vary in their strengths. The breaking stress of a thread or a wire (force per square centimetre) depends upon the material of which it is made. Stress is measured in *pascals* (Pa). 1 pascal = 1 newton per metre squared, or, more briefly, $1 \text{ Pa} = 1 \text{ N/m}^2$.

Breaking stresses

Attach a can to a cotton thread which is hanging from a support. Put small pebbles in the can, one at a time, until the cotton breaks. Now weigh the can and its contents. Repeat with threads of hair, nylon, wool and thin copper, steel and iron wires of the same gauge and compare their breaking stresses. The materials stretch before they break.

Building bridges

Perhaps you have noticed that the bars in metal bridges are arranged in triangular shapes. The reason for this is that rectangular shapes collapse easily whereas triangular shapes do not. A change in shape of a triangle is prevented by the tensions in its bars, which will not stretch and break unless tremendous stresses are applied. A change of shape of a rectangle occurs without any stretching or breaking of its bars.

A load on a humpbacked bridge is carried by the sides of its supports. A load on a flat bridge is carried by the tops of its supports, and there is a tendency for the centre of the bridge to bend or break.

Bridge structures

Use paper fasteners and card strips, 15 cm long and 2 cm wide, to make a rectangle and a triangle. Try to change their shapes. The shape of the rectangle is easily altered; the shape of the triangle cannot be altered. Can you see why?

Lay a postcard across two blocks of wood. Place a 100 gf weight on the card. The card bends downwards. Now bend the card upwards and hold it in place with pins. It supports the 100 gf weight without bending. Most of the weight is really supported by the pins at the sides and not by the card.

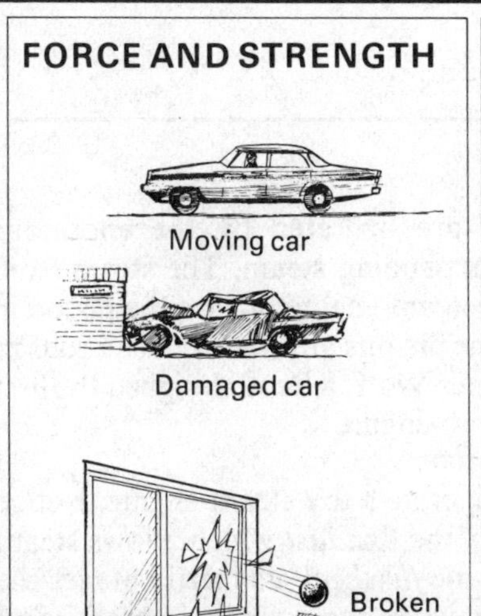
FORCE AND STRENGTH
Moving car
Damaged car
Broken window

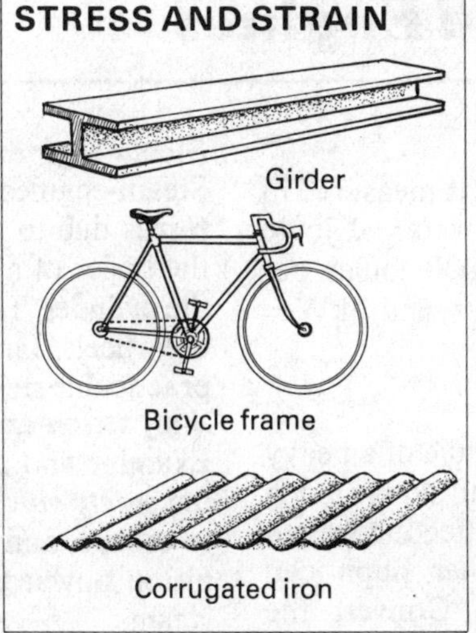
STRESS AND STRAIN
Girder
Bicycle frame
Corrugated iron

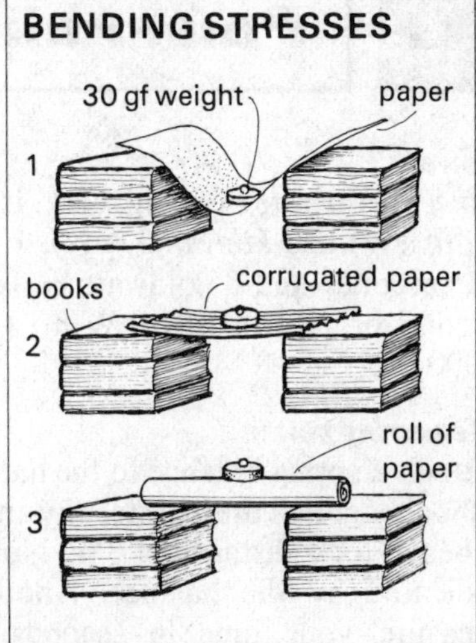
BENDING STRESSES
30 gf weight
paper
1
books
corrugated paper
2
roll of paper
3

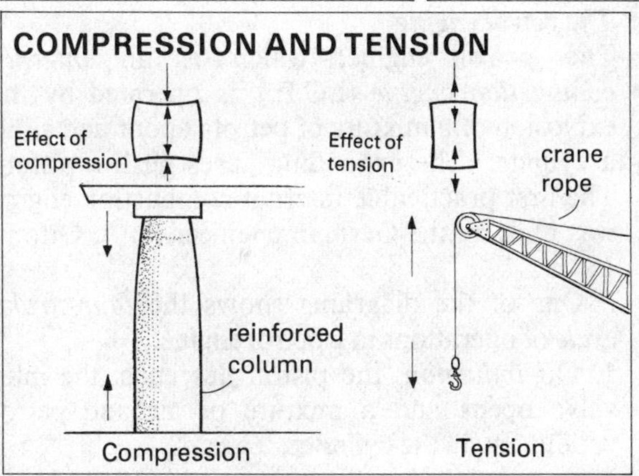
COMPRESSION AND TENSION
Effect of compression
Effect of tension
crane rope
reinforced column
Compression
Tension

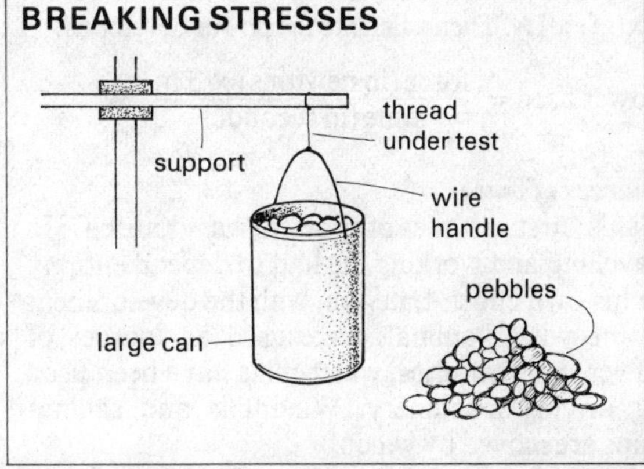
BREAKING STRESSES
thread under test
support
wire handle
pebbles
large can

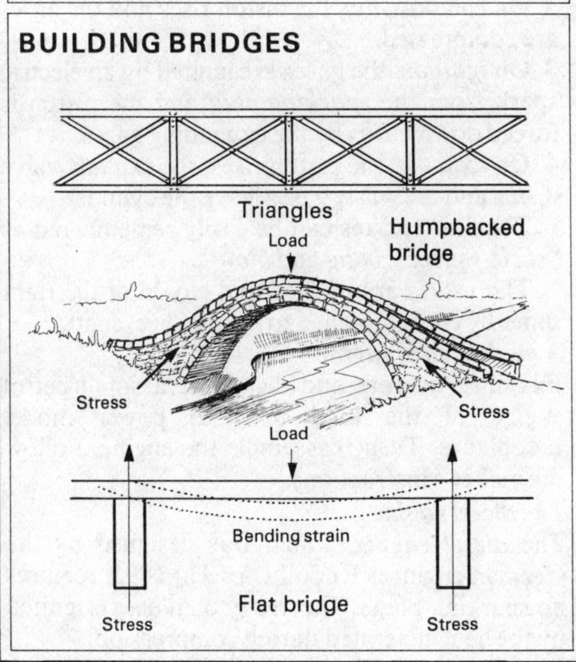
BUILDING BRIDGES
Triangles
Load
Humpbacked bridge
Stress
Stress
Load
Bending strain
Flat bridge
Stress
Stress

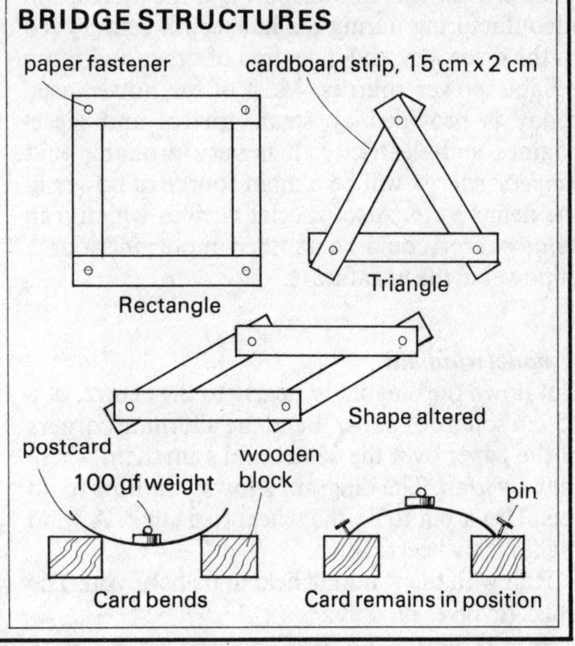
BRIDGE STRUCTURES
paper fastener
cardboard strip, 15 cm x 2 cm
Rectangle
Triangle
Shape altered
postcard
100 gf weight
wooden block
pin
Card bends
Card remains in position

23 Power and Engines

Power
Power is the rate of doing work. It is measured in *watts* (W) and *kilowatts* (kW). 1 watt = 1 joule per second, and 1 kilowatt = 1000 joules per second, or, more briefly, 1 W = 1 J/s, and 1 kW = 1000 J/s.

Measuring power
Attach a spring balance to the handle of a heavy school bag. Pull the bag steadily and slowly along a bench for a distance of 3 m. Notice the weight indicated by the balance. Another pupil can measure your time in seconds. Convert the weight indicated into newtons. 1 kgf $\simeq$ 10 N. 100 gf $\simeq$ 1N. Then calculate the power developed.

$$\text{Power used} = \frac{\text{force (in newtons)} \times 3\text{ m}}{\text{time (in seconds)}}\ \text{W}$$

Sources of power
Man's first source of power was himself. In travelling and working, he had to depend entirely on his own effort. Later on, with the development of the wheel, animals were used as sources of power. For centuries, water-mills have been used for driving machinery. Windmills and sailing-ships are moved by winds.

Improvements in transport and the increase in manufacturing during the nineteenth century led to the discovery and invention of other and more reliable power sources. Most of the power used today is provided by steam, petrol and diesel engines and electricity. It is very probable that *nuclear energy* will be a main source of power in the near future. Also, special devices which trap *solar energy* could be another important source of power in the near future.

A model windmill
Cut down the diagonals, nearly to the centre, of a 15 cm square of paper. Bend the alternate corners of the paper over the centre and gum them where they overlap. The diagram shows you how to do this. Use a pin to fix the wheel to a stick. A wind makes the wheel revolve.

Run with the windmill held in front of you. The wheel revolves quickly.

Steam-engines
Steam-engines are operated by the enormous forces due to expanding steam. The steam from the boiler of a steam-engine pushes the piston in the cylinder. The pistons are connected by rods to the wheels. James Watt, a Scot, designed the first practicable steam-engine.

A toy steam-engine
Examine and operate a toy steam-engine. Notice the *safety-valve,* the *exhaust* which allows steam to escape, and the *flywheel;* its inertia moves the piston upwards when the cylinder is emptied of steam.

The petrol engine
The petrol engine, which is an *internal combustion engine* (I.C.E.), is operated by the explosion of a mixture of petrol vapour and air in a cylinder. The expanding gases push a piston. The first practicable internal combustion engine was built by the German engineer N. A. Otto in 1876.

One of the diagrams shows the *four-stroke cycle* of operations in a petrol engine.

1. On *induction,* the piston descends, the inlet valve opens and a mixture of air and petrol vapour enters the cylinder.
2. On *compression,* the piston rises and the gases are compressed.
3. On *ignition,* the gases are ignited by an electric spark from the *sparking-plug* and the piston is forced downwards by the expanding gases.
4. On *exhaust,* the piston rises, the exhaust valve opens and the waste gases leave the cylinder.

The four strokes can be easily remembered as "*suck, squeeze, bang* and *blow*".

The valves are opened and closed at the right times by *cams* attached to the flywheel shaft.

A model petrol engine
Examine, operate and dismantle a small petrol engine of the kind used to power model aeroplanes. Then reassemble the engine. Follow the maker's instructions.

The diesel engine
The *diesel engine,* which was designed by the German engineer Rudolf Diesel in 1895, requires no sparking-plugs. The heavy oil it uses is ignited by the heat generated during compression.

MEASURING POWER

$$\text{Power} = \frac{\text{work}}{\text{time}}$$

MODEL WINDMILL

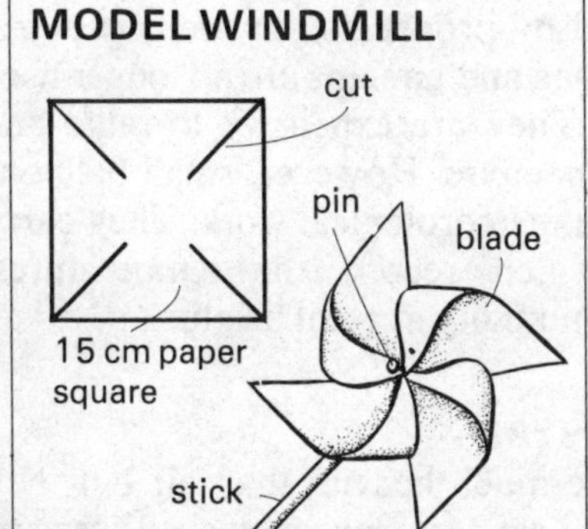

THE FOUR-STROKE CYCLE

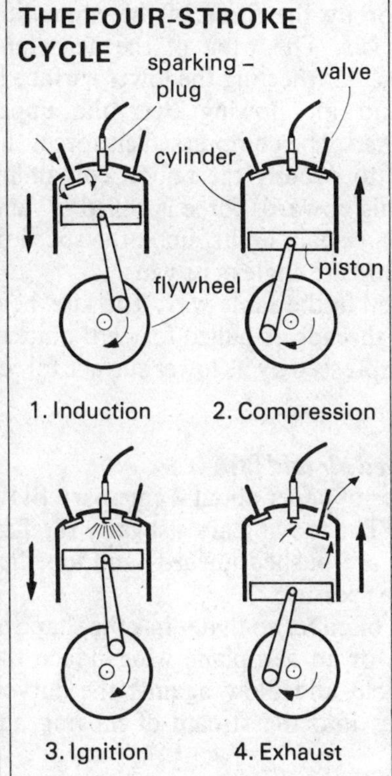

SOURCES OF POWER

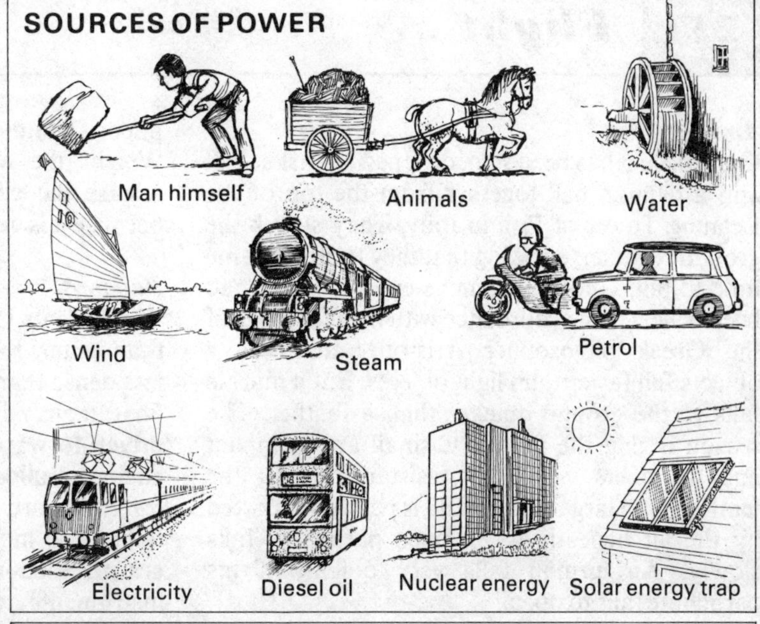

TOY STEAM-ENGINE

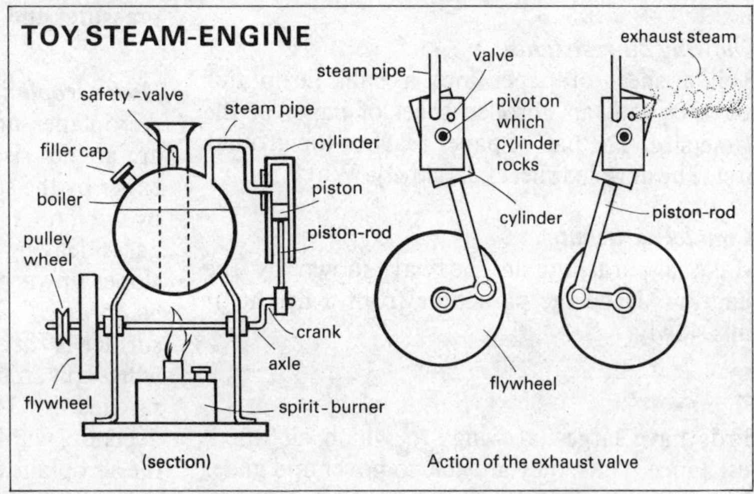

JAMES WATT
1736-1819

Designer of steam-engines

PETROL ENGINE

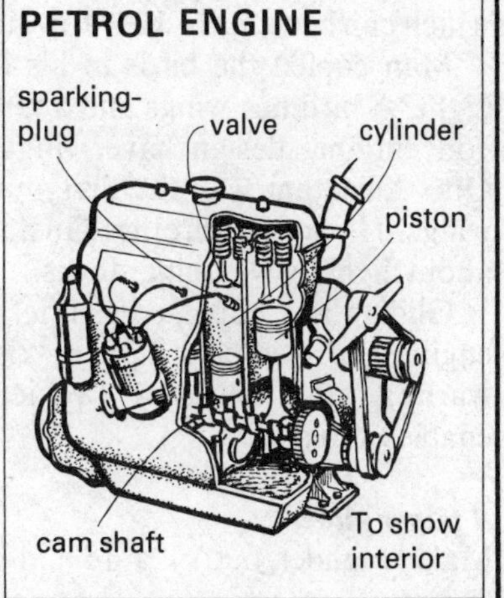

24 Flight

Air resistance
Galileo, so it has been said, dropped a musket ball and a cannon ball together from the top of the Leaning Tower of Pisa in Italy. They struck the ground together, showing that they took the same time to fall. Until this famous experiment, it had been believed, in accordance with the teachings of the Greek philosopher Aristotle, that heavy objects fall faster than light objects. But a marble falls to the ground quicker than a feather. The reason is that the marble is small and compact and air offers very little resistance to it; the feather has a large surface and is partly supported by the air beneath it. An open parachute falls slowly. An airman falls very quickly if his parachute fails to open.

Showing air resistance
Roll a sheet of paper into a ball. Drop this together with an unrolled sheet of paper of the same size. The ball of paper reaches the ground first. The unrolled sheet falls slowly. Why?

A model parachute
Make a parachute in the way shown by the diagram. Drop the parachute from a height. It falls slowly.

Flight
Birds have large flat wings to which air offers resistance. Thus, they are able to hover and glide. Birds move by flapping their wings in a manner which can be described as "rowing in air".

Man copied the birds in his first attempts at flight. A bird has wings and a tail. Aeroplanes of conventional design have wings and tails also. Otto Lilienthal was the first man to glide with wings. He and his brother Gustav learned much about flight by watching storks.

Gliding is a popular pastime. A glider has no engine and it is held aloft by rising currents of warm air. It is towed by a vehicle until its speed enables it to rise.

A paper glider
Make a glider out of a 15 cm square of thick paper. The diagrams show you how to do this. Weight the nose with a paper-clip. Jerk and release the glider. It will fly for some distance because it is held aloft by air resistance.

Balloons
Wood floats in water because wood is less dense than water. Balloons, which are filled with gases less dense than air, such as hydrogen and helium, float in air. Airships are rigid balloons which are driven forward by propellers powered by petrol engines. Balloons and airships are no longer used for transport. They are expensive to build and difficult to manoeuvre. However, small balloons are still used in meteorological work. They carry instruments which record the temperatures, pressures and humidity at great heights.

How aeroplanes rise
Aeroplanes are much heavier than air but they are able to rise. An aeroplane is pushed forward either by the stream of rapidly moving air created by *airscrews* or by the force of the expanding gases leaving *jets*. The wing of the aeroplane slopes upwards. Air meeting the lower surface is *compressed* and air flowing over the upper surface is *rarefied*. The compressed air forces the wing upwards to occupy the region containing rarefied air. This upwards force is called *lift* and depends, within certain limits, upon the speed of the aeroplane and the angle of its wing.

A kite is lifted in the same way. The kite, held at an angle by threads, is pulled forward quickly and the air compressed by its lower surface forces it to rise.

Showing rarefied air and lift
Hold two sheets of paper about 2 cm apart. Blow between them. The moving air is slightly rarefied and the sheets are pushed inwards and together by atmospheric pressure.

Fold a sheet of paper so that it has the shape of the upper side of an aeroplane wing. Place the paper on a table and blow against the curved surface. It rises into the stream of moving air. Why?

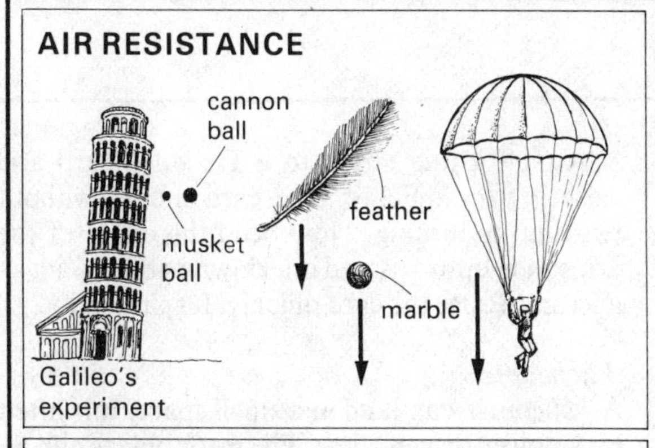
AIR RESISTANCE
cannon ball
musket ball
feather
marble
Galileo's experiment

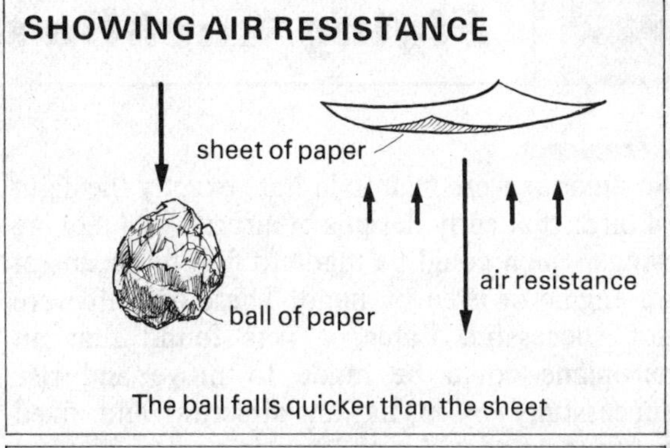
SHOWING AIR RESISTANCE
sheet of paper
air resistance
ball of paper
The ball falls quicker than the sheet

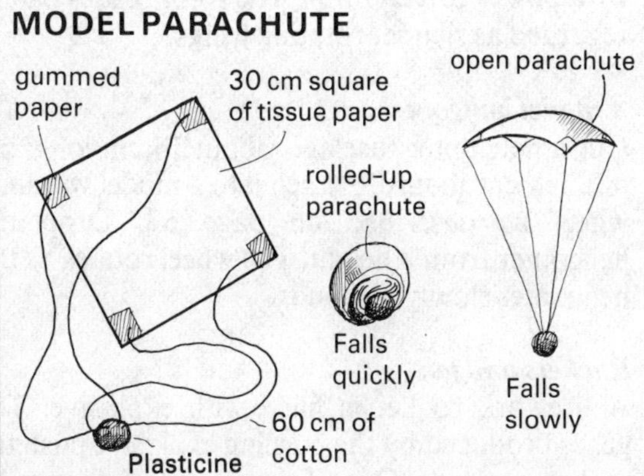
MODEL PARACHUTE
gummed paper
30 cm square of tissue paper
open parachute
rolled-up parachute
Falls quickly
Falls slowly
60 cm of cotton
Plasticine

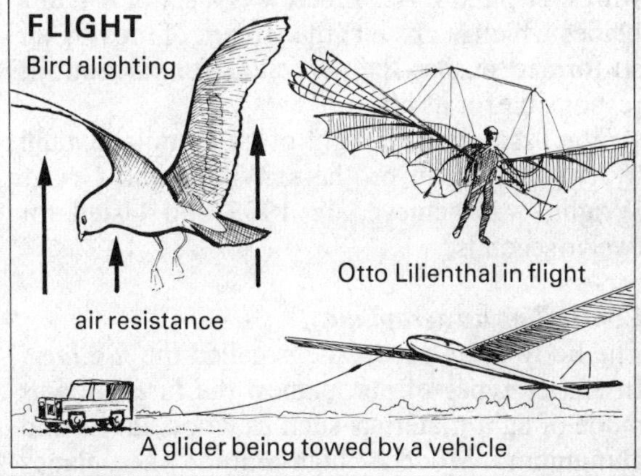
FLIGHT
Bird alighting
Otto Lilienthal in flight
air resistance
A glider being towed by a vehicle

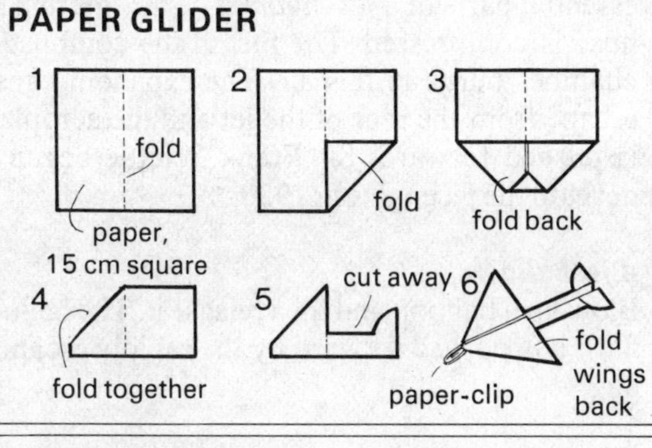
PAPER GLIDER
1
2
3
fold
fold
fold back
paper, 15 cm square
4
5
cut away
6
fold together
paper-clip
fold wings back

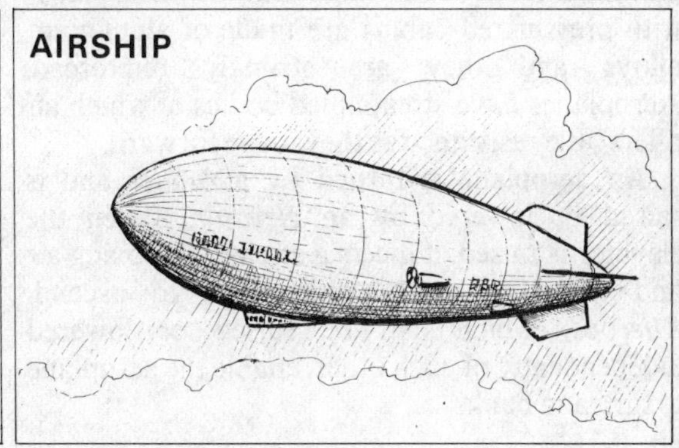
AIRSHIP

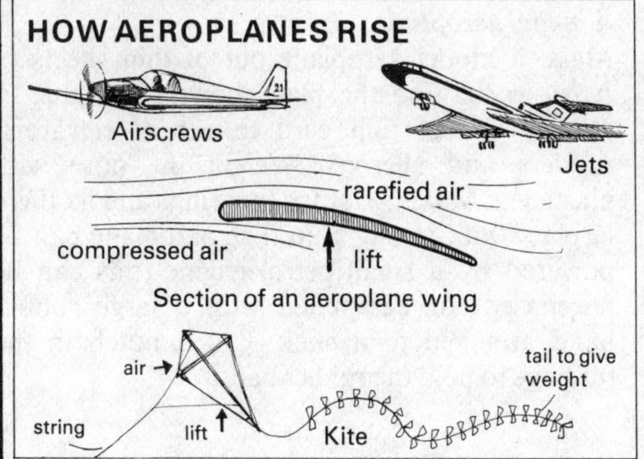
HOW AEROPLANES RISE
Airscrews
Jets
rarefied air
compressed air
lift
Section of an aeroplane wing
air
tail to give weight
string
lift
Kite

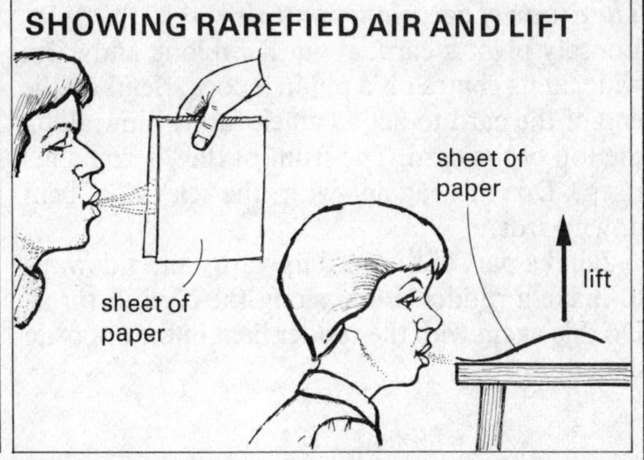
SHOWING RAREFIED AIR AND LIFT
sheet of paper
lift
sheet of paper

25 Flying Machines

Aeroplanes
So anxious were men to imitate exactly the flight of birds that early designs of aircraft had moving wings which could be made to flap by means of an engine or even by hand. These aircraft were not successful. Later, it was found that an aeroplane could be made to move and rise successfully by using an airscrew and fixed wings, or *planes*. An airscrew is a set of rotating blades which sucks air; the stream of moving air so formed pushes the aeroplane forward and is responsible for its lift.

The first powered flight of an aeroplane, built by two American brothers, Wilbur and Orville Wright, was achieved in 1903 and lasted for twelve seconds.

Controlling an aeroplane
The body of an aeroplane is called the *fuselage*. In earlier types of aeroplanes, the fuselage was made of light materials such as wood, fabric and aluminium. Modern high-altitude aeroplanes with pressurized cabins are made of aluminium alloys and they are strongly reinforced. Aeroplanes have streamlined bodies to which air offers little resistance as they move forward.

An aeroplane is turned by a *rudder* and is raised or lowered by an *elevator*. When the elevator is raised, it meets a stream of moving air and the aeroplane is compelled to ascend. *Ailerons,* which can be raised or lowered independently of each other, enable the aeroplane to turn and bank.

The action of aeroplane controls
Loosely pivot a card, about 8 cm long and 2 cm wide, at its centre on a pin in a cork. Bend up the end of the card to act as an elevator. Blow along the top of the card. The front of the "aeroplane" rises. Do this again with the elevator bent downwards.

Bend a part of the card upwards and sideways to make a rudder. Blow along the card. It turns. Do this again with the rudder bent in the opposite direction.

Attach a thin string to a T-shaped card and raise it very quickly. The card moves without twisting or turning. Now bend the edges of the arms, one upwards and one downwards, to act as ailerons. Raise the card quickly. It spins. Why?

A helicopter
A *helicopter* can land in a small space. It is lifted by a power-driven *rotor*. The rotor blades can be regarded as slender rotating wings.

A model helicopter
Cut a helicopter fuselage, about 15 cm long, out of thick cardboard. Attach it to a model windmill wheel as described on page 52. Drop the helicopter from a height. The wheel rotates as the helicopter slowly descends.

Rockets and jets
A firework rocket is filled with explosive. The gases produced by the burning explosive push the rocket upwards. One of the diagrams shows the essential parts of a *jet engine*. The air entering the nose is compressed. The fuel in the combustion chamber burns in this air. The expanding gases escape from the rear of the jet and the aeroplane is pushed forward. Sir Frank Whittle began to develop the jet engine in 1930.

A jet balloon
Blow up a balloon and then release it. The balloon flies; it is pushed forward by the rapidly escaping air.

A model aeroplane
Make a model aeroplane out of thin sheets of *balsa-wood.* Use the plan shown. Use glue to attach tabs of thin card to act as elevators, rudders and ailerons. Weight the nose with plasticine. Make a slot for the wings and fix them in place with strong glue. The aeroplane can be powered by a small petrol engine (this can be purchased) or catapulted with a large rubber band attached to a stick. Cut a notch in the fuselage to hold the rubber band.

EARLY AEROPLANE

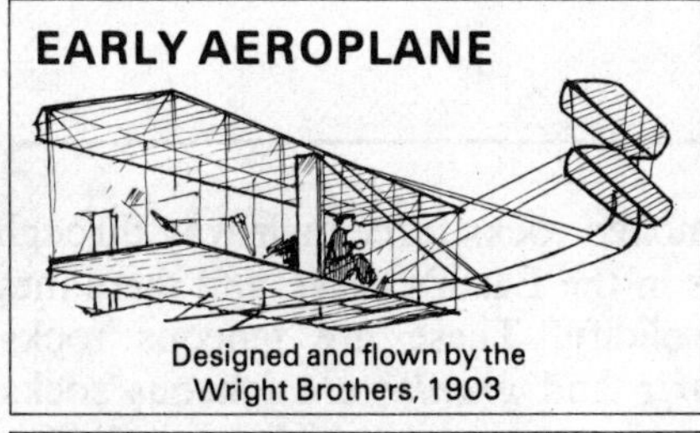

Designed and flown by the Wright Brothers, 1903

ACTION OF AEROPLANE CONTROLS

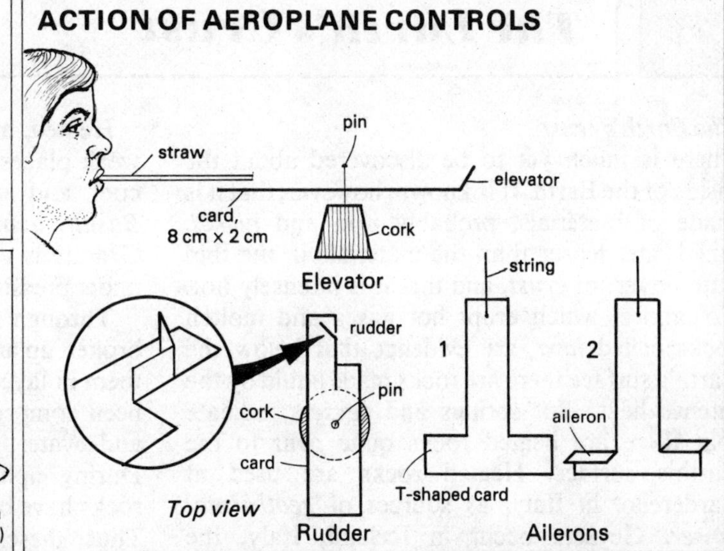

CONTROLLING AN AEROPLANE

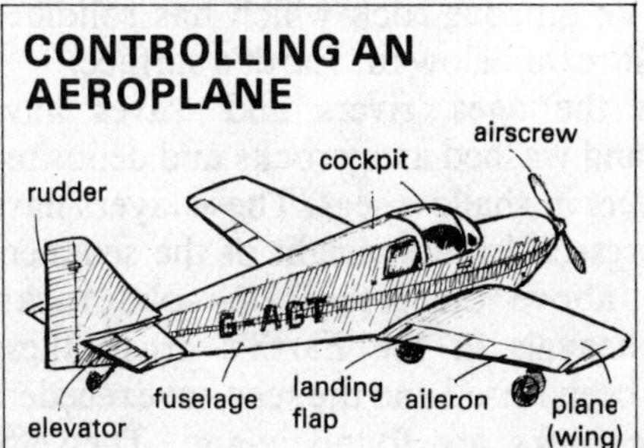

HELICOPTER

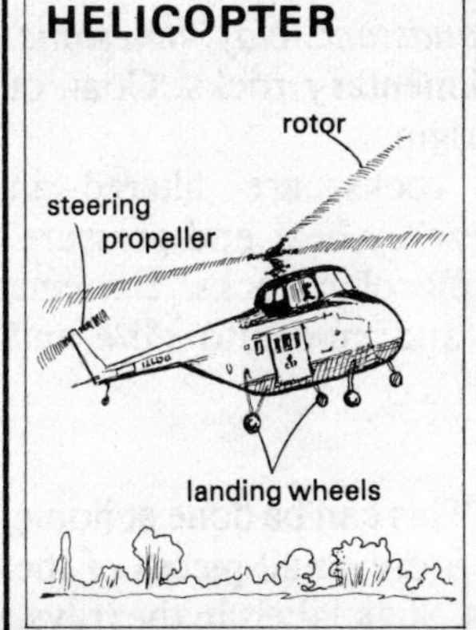

MODEL HELICOPTER

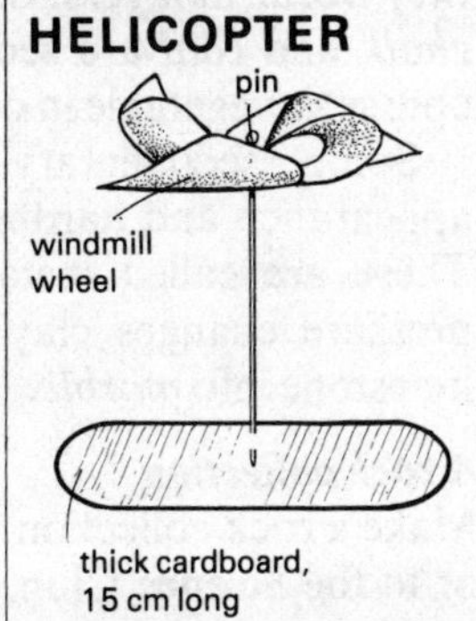

SIR FRANK WHITTLE

Began to develop the jet engine in 1930

JET BALLOON

HOW ROCKETS WORK

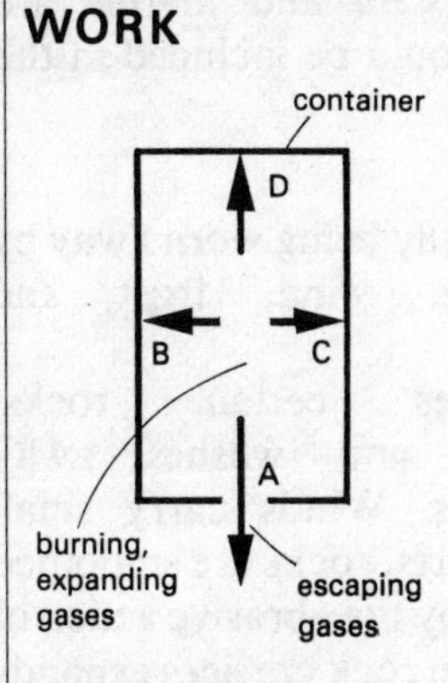

The forces of the expanding gases on the sides B and C counter-balance each other. The force against the front, D, of the container pushes the rocket forward. The gases merely escape at A.

ROCKETS AND JETS

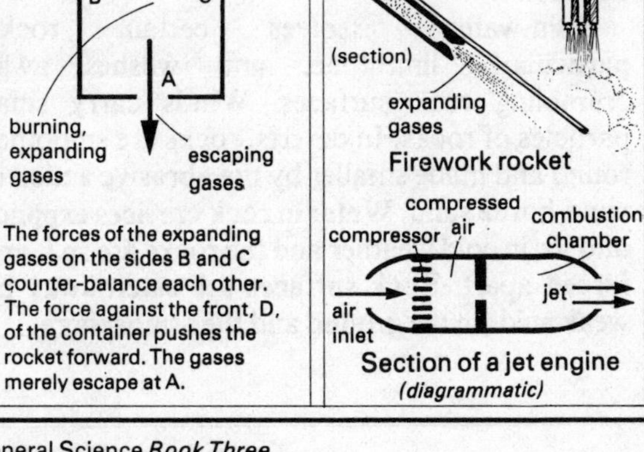

Section of a jet engine
(diagrammatic)

MODEL AEROPLANE

Use balsa-wood, 3 mm thick for fuselage and 2 mm thick for wings and tail

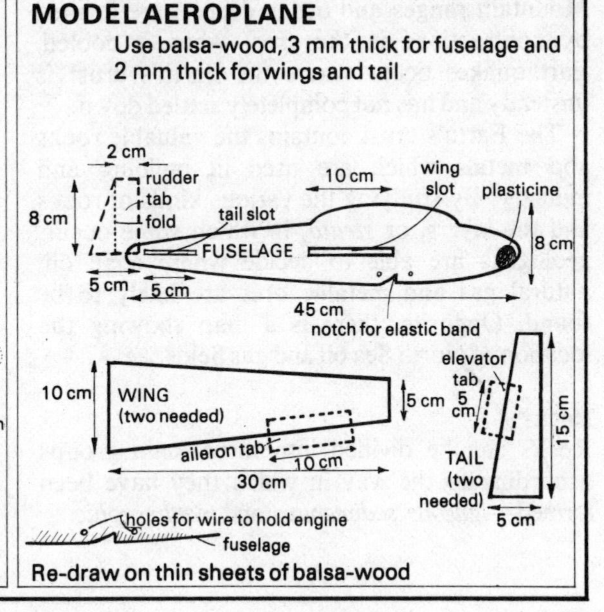

Re-draw on thin sheets of balsa-wood

26 The Earth's Crust

The Earth's crust

There is much yet to be discovered about the inside of the Earth. It is known, however, that it is made of materials, probably iron and *nickel*, which are denser than the materials in the thin outer layer, or *crust*, and that it is intensely hot. Volcanoes, which erupt hot gases and molten rocks called *lava*, are evidence that below the Earth's surface there are rocks made liquid by the intense heat. Hot springs and *geysers* indicate that there are heated rocks quite near to the Earth's surface. Heated rocks are used at Larderello, in Italy, as sources of *geothermal power*. Geysers occur in Iceland, Italy, the Yellowstone National Park region of Wyoming in the United States, and North Island, New Zealand.

It is believed that the Earth was once a ball of burning gases pulled out of the Sun by the gravitational force of a passing star. On cooling—a process which has taken millions of years—the gases and vapours condensed to form liquids. Some oxygen and some nitrogen have remained as gases in the atmosphere. Water vapour, formed by hydrogen burning in oxygen, condensed to give the seas and oceans. After further cooling, the liquid materials near the Earth's surface solidified to form an outer crust. Mountain ranges and ocean deeps were formed by contractions in the crust when it cooled. Earthquakes occur where the Earth's crust is unsteady and has not completely settled down.

The Earth's crust contains the valuable rocks and metals which are used in building and industry. By studying the various kinds of rocks and the layers, or *strata*, in which some occur, geologists are able to decide where coal, oil, natural gas and metallic ores are likely to be found. Opposite, there is a map showing the location of North Sea oil and gas fields.

Rocks

Rocks can be divided into three main groups according to the way in which they have been formed—*igneous, sedimentary* and *metamorphic*.

Heated, molten rocks force their way through weak places in the Earth's crust, and there they cool and solidify. These are igneous rocks. *Basalt, quartz* and *granite* are igneous rocks. Granite is an igneous rock which has solidified under pressure far below the Earth's surface.

Through the ages, rivers and waves have broken up and washed away rocks and deposited them in layers in shallow seas. These layers have been compressed by the weight of the sediment and water above them to form solid rocks. During upheavals in the Earth's crust, these rocks have been raised and the seas have receded. Thus, these rocks are found inland. They are called sedimentary, or *stratified*, rocks because they occur in layers. *Sandstone, clay, limestone, chalk* and *coal* are sedimentary rocks. Coal, of course, is vegetable in origin.

Some sedimentary rocks are altered in appearance and hardness by heat and pressure. These are called metamorphic rocks. Extreme pressure changes clay and mud into *slate* and limestone into *marble*.

A rock collection

Make a rock collection. This can be done at home or in the Science Club. Store small pieces of the rocks in matchbox trays. Stick labels in the trays. Granite, basalt, quartz, *pumice*, sandstone, clay, limestone, chalk, coal, slate and marble are common rocks which should be included in the collection.

Weathering

Rock surfaces are gradually being worn away by *weathering agents*—rain, wind, frost and chemicals.

Rain-water dissolves certain rocks, particularly limestone, and washes away crumbling rock surfaces. Winds carry small particles of rocks. In deserts, rocks are smoothed round and made smaller by the abrasive action of wind-borne sand. Water in rock crevices expands into ice in cold weather and the rocks are split and forced apart. Rock surfaces are eaten away by weak acids in the ground and the atmosphere.

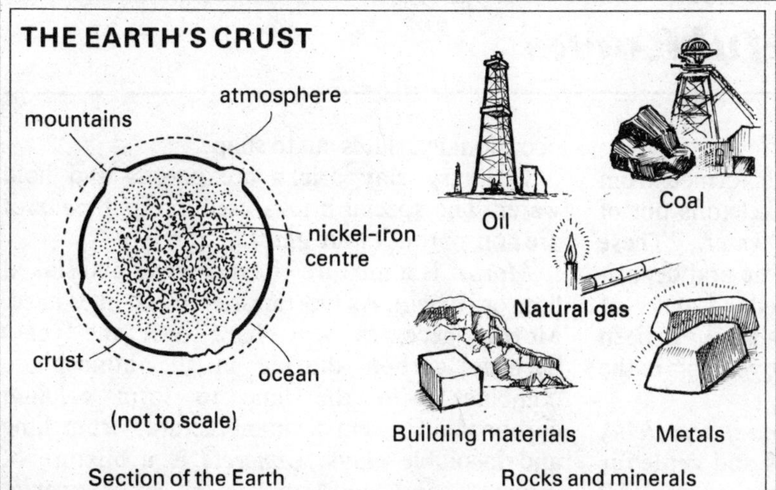
THE EARTH'S CRUST
atmosphere
mountains
nickel-iron centre
crust
ocean
(not to scale)
Section of the Earth
Oil
Coal
Natural gas
Building materials
Metals
Rocks and minerals

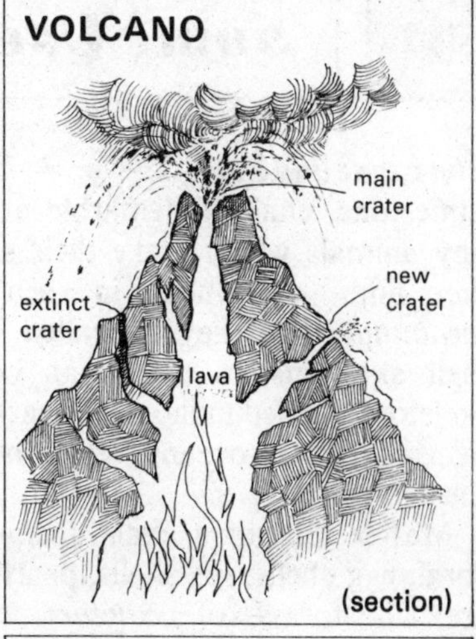
VOLCANO
main crater
new crater
extinct crater
lava
(section)

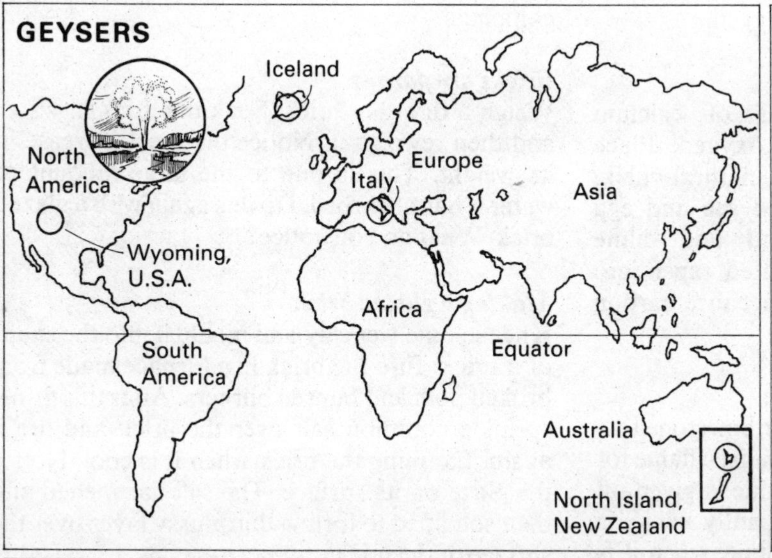
GEYSERS
Iceland
North America
Europe
Italy
Asia
Wyoming, U.S.A.
Africa
South America
Equator
Australia
North Island, New Zealand

NORTH SEA OIL AND GAS
North Sea
GREAT BRITAIN
Oil and gas field

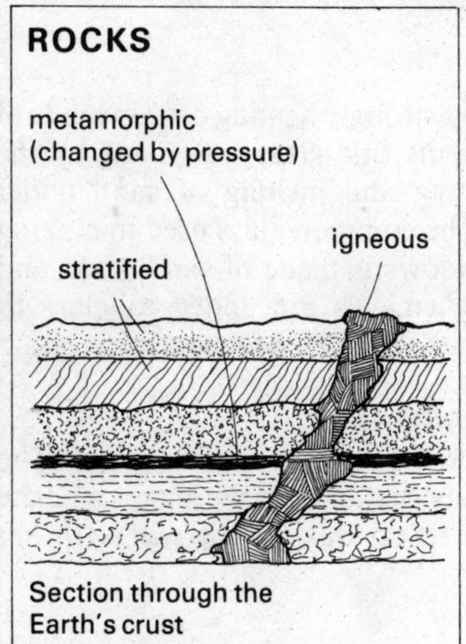
ROCKS
metamorphic (changed by pressure)
igneous
stratified
Section through the Earth's crust

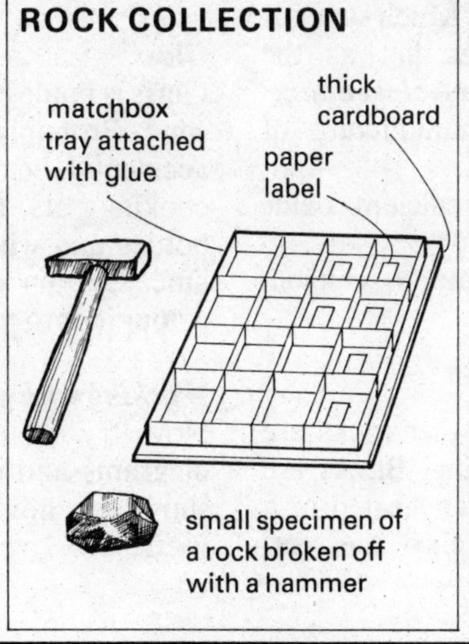
ROCK COLLECTION
thick cardboard
matchbox tray attached with glue
paper label
small specimen of a rock broken off with a hammer

WEATHERING
Rain
Frost
Wind
Chemical action

27 Some Useful Rocks

Limestone and chalk
Limestone, chalk and marble are formed from tiny animals which make their skeletons out of the lime contained in sea-water. These *foraminifera,* as they are called, die and deposit their skeletons on the ocean beds. Layers of skeletons formed millions of years ago have been raised above the ocean level by movements in the Earth's crust.

Marble, limestone, chalk, *Iceland spar, calcite,* coral, egg shells, sea shells, pearls and kettle fur are all forms of *calcium carbonate.* All carbonates contain carbon and oxygen.

Testing for calcium carbonate
Show that the various forms of calcium carbonate contain carbon and oxygen. Place small quantities of marble chips, natural chalk, limestone, kettle fur and crushed sea and egg shells in separate beakers and add dilute hydrochloric acid. Lower a lighted taper into each beaker. It is extinguished because carbon dioxide gas has been formed.

Making quicklime and slaked lime
Attach a lump of natural chalk or limestone to a wire and strongly heat it in a hot bunsen flame for at least 15 minutes. Carbon dioxide is given off and *quicklime,* which glows brilliantly white, is formed. Place some of the quicklime, when it is cool, in a can. Drip water on to it. Much steam is produced and the can becomes hot as the quicklime absorbs water to become *slaked lime.* Slaked lime is used in the manufacture of *bleaching powder.*
Calcium carbonate (chalk) = calcium oxide (quicklime) + carbon dioxide
Calcium oxide (quicklime) + water = calcium hydroxide (slaked lime)

Building materials
Sandstone, granite, marble and other rocks are cut to shape and used for building. Bricks are moulded out of wet clay and then heated in a furnace. Roofs are built of clay tiles or, occasionally, slates cut to shape.

Ordinary clay bricks are porous and hold water. The special bricks used in *damp courses* are non-porous or are *glazed.*

Mortar is a mixture of sand, water and slaked lime or *cement.* As mortar dries, it becomes hard. Mortar becomes very hard over the years because carbon dioxide in the atmosphere combines with the lime to form calcium carbonate. Cement is manufactured from lime and insoluble clays. *Concrete* is a mixture of cement, sand and small pebbles or granite chippings.

Bricks are porous
Weigh a dry clay brick. Soak the brick in water and then reweigh it. Notice the great increase in its weight. This is due to the large amount of water it has absorbed. Do this again with a glazed brick. What do you notice?

Making a glazed brick
Knead some wet clay and mould it into the shape of a brick. Fire the brick in a furnace made from broken pots and bunsen burners. After this firing, sprinkle common salt over the brick and fire it again. Examine the brick when it is cool. Notice the glaze on its surface. The salt has melted and then solidified to form a thin glassy layer over the surface of the brick.

Glass
Glass is made by strongly heating certain kinds of sand. Probably the first glass was made by the accidental heating and melting of sand under cooking pots. The common glass used in making bottles and windows is made of sand, soda and lime. Certain chemicals are added to glass to colour it or to make it suitable for special work.

Working with glass
Now try to do some work with glass. The diagrams and instructions on the page opposite show you how to do this. *Be careful with hot glass!*

FORAMINIFERA

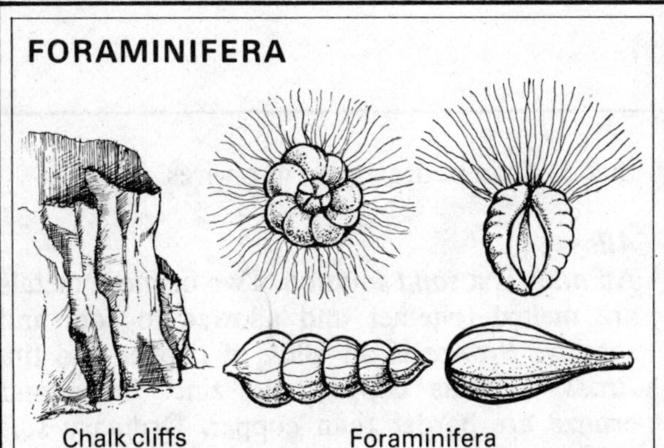

TESTING FOR CALCIUM CARBONATE

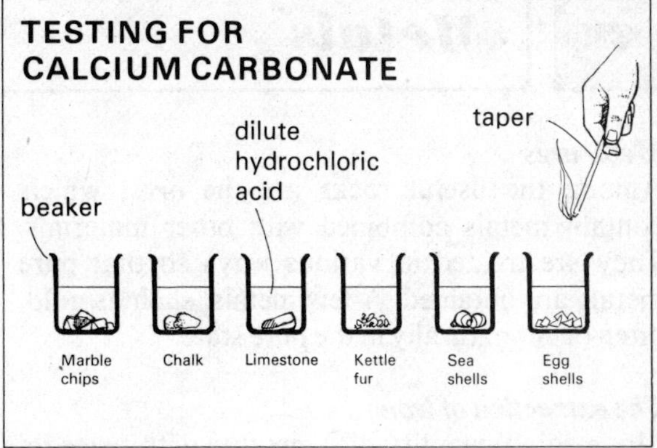

MAKING QUICKLIME AND SLAKED LIME

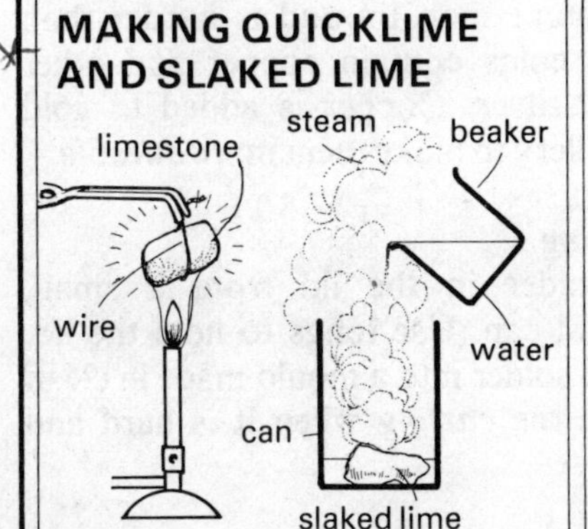

BUILDING MATERIALS

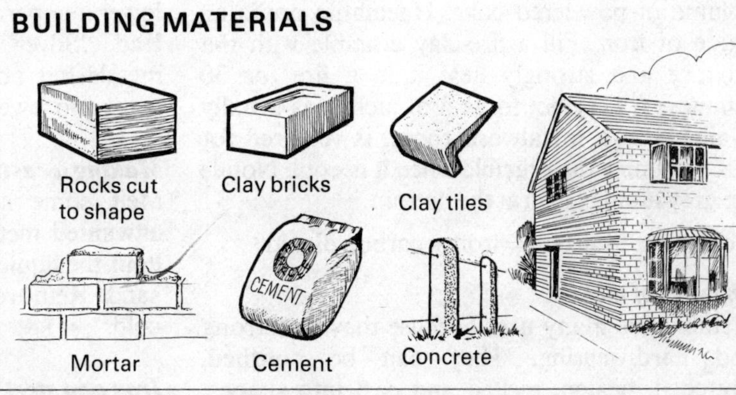

BRICKS ARE POROUS

WORKING WITH GLASS

1. Cutting

Scratch a mark about 20 cm from the end of a glass tube. Tap the tube sharply. The tube breaks at the mark.

2. Fire-polishing

Rotate one end of the tube in a bunsen flame. The end of the tube melts and becomes smooth.

3. Sealing

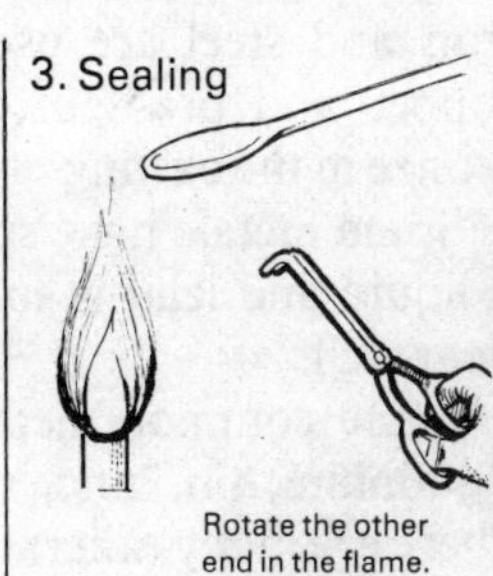

Rotate the other end in the flame. When the glass is soft, seal by squeezing with tongs.

4. Bending

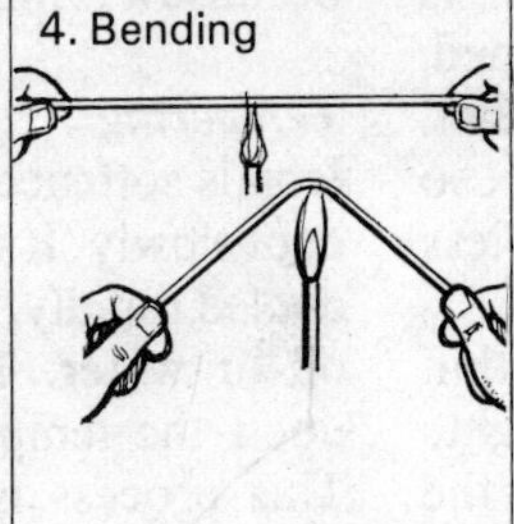

Rotate the middle of the tube in the flame. Bend the tube to make a right angle.

5. Stretching

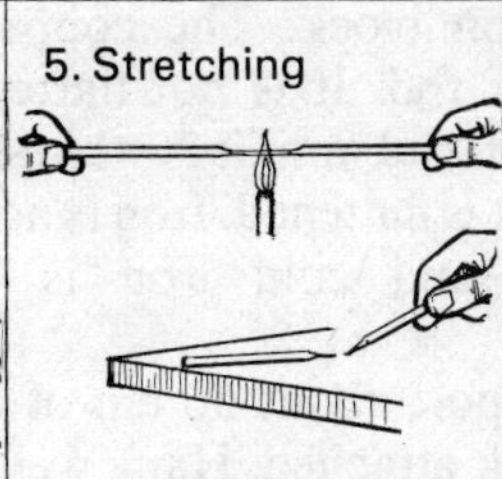

Rotate the middle of another tube in the flame. Stretch the tube by pulling. When cool, snap it at its middle to make two jets.

6. Making a bulb

Heat another tube for about 2 cm along its end. When the end is soft and sealed, blow down the other end to make a bulb.

MAKING A GLAZED BRICK

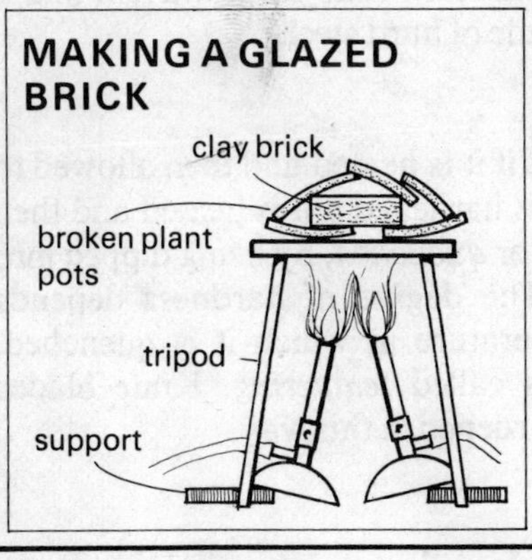

28 Metals

Metal ores
Among the useful rocks are the *ores*, which contain metals combined with other materials. They are treated in various ways so that pure metals are obtained. A few metals, such as gold, often occur naturally in the pure state.

The extraction of iron
Mix a small quantity of *haematite* with twice its volume of powdered coke. Haematite ore is an oxide of iron. Fill a fire-clay crucible with the mixture and strongly heat it in a fire for 30 minutes. A very hot forge fire, such as is usually available in a metalwork room, is required for this. Examine the crucible when it is cool. Notice the grey lump of iron at the bottom.

Iron oxide + carbon = iron + carbon dioxide

The uses of metals
Metals have many uses because they are strong and hard-wearing. They can be polished, stretched, beaten, melted and cast into shapes. They are good conductors of heat and electricity. Iron and steel are used in making machinery, copper and brass in electrical equipment, and bronze in the casting of ornaments.

Some metals have special properties. Mercury is liquid and lead is soft. Gold and silver do not wear well.

Some common metals are copper, iron, lead, aluminium, tin, zinc, magnesium and mercury. There are many others.

Some properties of metals
Use a hammer to beat a short length of thick copper wire on an iron block. The copper is flattened. Beat an iron rod. It is not flattened. Heat the end of the iron rod until it is white-hot. Now beat it again and it is flattened. Iron is not so *malleable* as copper and cold iron is less malleable than hot iron.

Suspend from a support about 50 cm of thin copper wire with a hook attached. Hang weights on the hook until the wire breaks. Notice that the wire stretches before it breaks. Copper is *ductile,* that is, it can be drawn out into wires.

Alloys
An *alloy* is a *solid solution.* Two or more metals are melted together and allowed to cool and solidify. *Bronze* is an alloy of copper and tin. *Brass* contains copper and zinc. Brass and bronze are harder than copper. Ordinary *soft solder* is a mixture of tin and lead. It melts at a lower temperature than tin and is harder than lead. "Silver" coins contain copper and other metals but no silver. Copper is added to gold coins and jewellery to make them more *durable.*

Making a casting
Melt some solder in the lid from a small, unwanted metal can. Use tongs to hold the lid. Pour the liquid solder into a mould made in clean sand. Remove the *casting* when it is hard and cold.

Iron and steel
The *cast iron* from a foundry blast-furnace contains many impurities that include carbon. It is hard and brittle and, therefore, it breaks easily. *Wrought iron* contains little or no carbon. It bends and, when heated, it is easily worked. *Steel* is made by adding carbon to iron. The very hard steels that are used in the manufacture of machine tools contain as much as 1.5% carbon.

Soft iron and hard steel
Use two pairs of pliers to bend a pin. The pin bends because it is made of soft iron. Now try to bend a sewing needle in the same way. It snaps because it is made of hard steel.

Tempering
Iron is softened if it is heated and then allowed to cool slowly. It is hardened if it is heated and then cooled rapidly, or *quenched,* by being dipped into oil or water. The degree of hardness depends upon the temperature at which it is quenched. This process is called *tempering.* Knife blades and tools are hardened in this way.

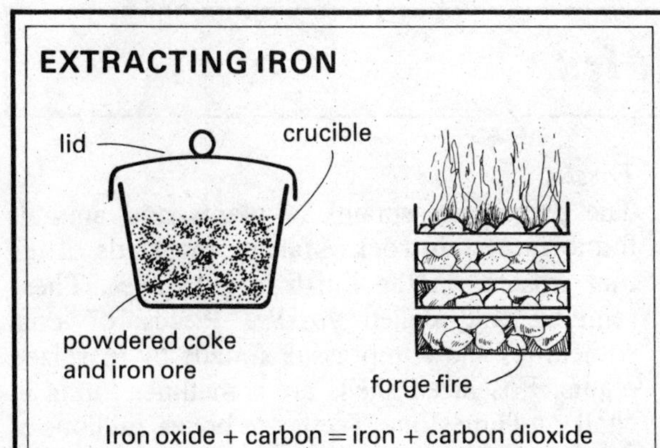
EXTRACTING IRON
lid
crucible
powdered coke
and iron ore
forge fire
Iron oxide + carbon = iron + carbon dioxide

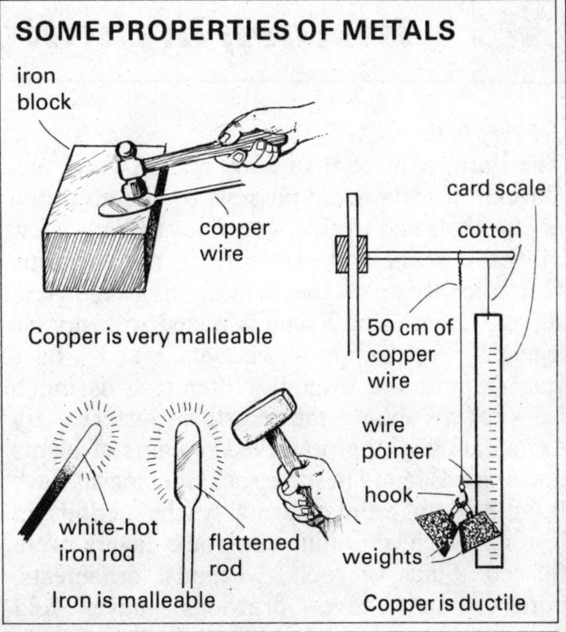
SOME PROPERTIES OF METALS
iron
block
copper
wire
Copper is very malleable
card scale
cotton
50 cm of
copper
wire
wire
pointer
hook
weights
Copper is ductile
white-hot
iron rod
flattened
rod
Iron is malleable

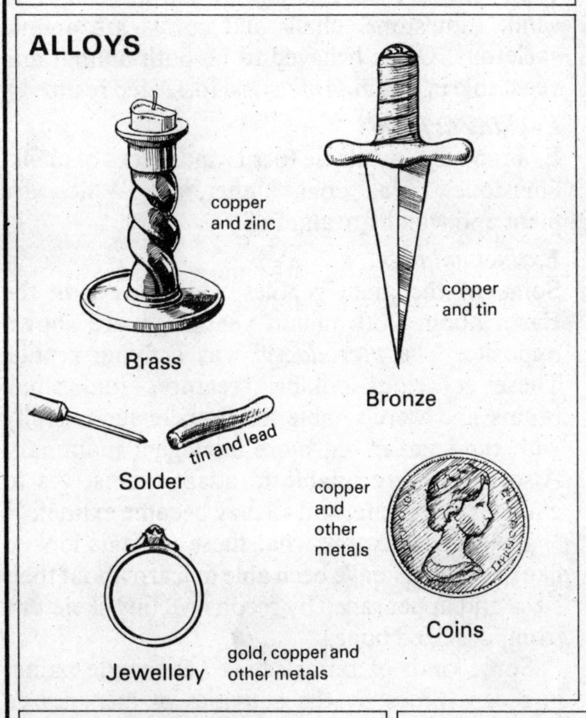
ALLOYS
copper
and zinc
Brass
copper
and tin
Bronze
tin and lead
Solder
copper
and
other
metals
Coins
Jewellery
gold, copper and
other metals

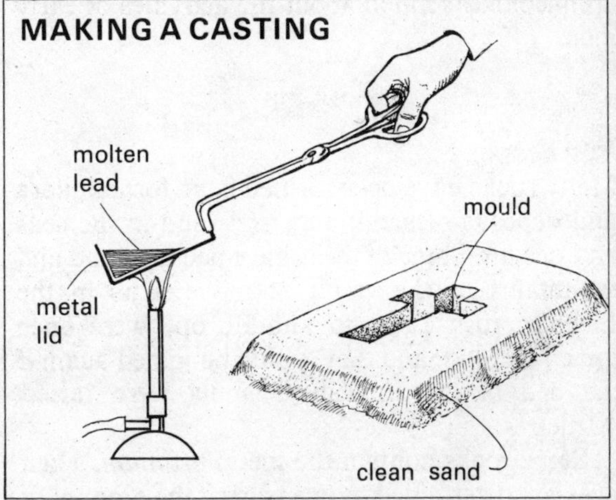
MAKING A CASTING
molten
lead
metal
lid
mould
clean sand

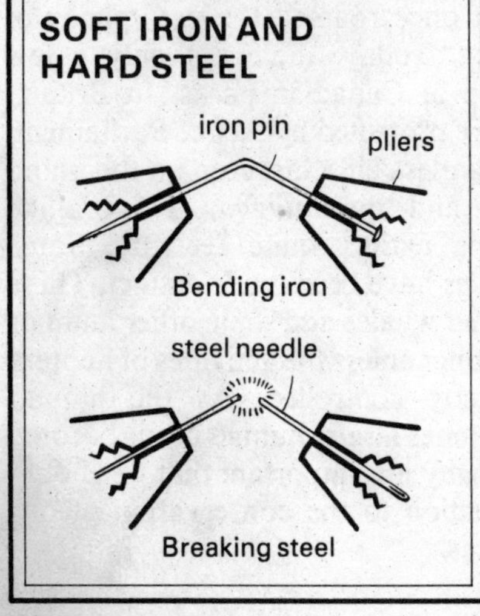
SOFT IRON AND
HARD STEEL
iron pin
pliers
Bending iron
steel needle
Breaking steel

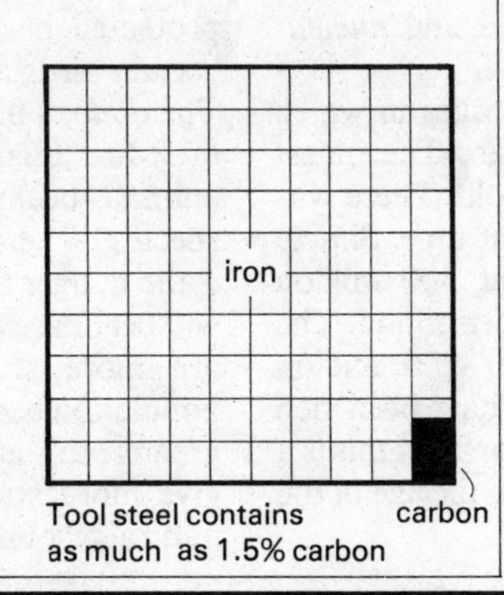
CARBON IN STEEL
iron
carbon
Tool steel contains
as much as 1.5% carbon

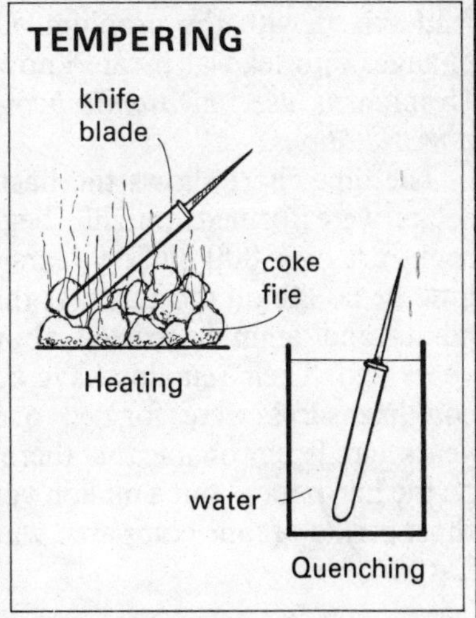
TEMPERING
knife
blade
coke
fire
Heating
water
Quenching

29 History in the Rocks

History in the rocks

The Earth is more than 2000 million years old. Written history, made possible by the invention of alphabets and writing, goes back for only a few thousand years, and some of it is not accurate. Our knowledge of the ancient civilisations of Egypt, Greece and Rome is mixed with myths, legends, conflicting accounts and false speculations. Yet, without written records, much is known about the Earth's history. By examination of the preserved remains of plants and animals found in some rocks, geologists have been able to estimate roughly the periods in history when the different rock layers were formed. Finds of tools, weapons, ornaments, pottery and cave drawings have told archaeologists much about the activities of early men.

Past ages

Many rocks have been formed from foraminifera and deposits of sand, clay and mud in the seas and oceans. Some of these have become land and mountains as the result of movements in the Earth's crust. England and Europe were once joined by land, and they would be joined again if the bed of the English Channel were raised 40 metres.

Some rocks contain the metal *uranium.* Their ages are determined by measuring the amount of lead mixed with the uranium. Uranium slowly changes into lead at a rate known to scientists. Uranium is used in *atomic bombs* and *nuclear power-stations.*

The time-chart shows the past ages in which rocks were formed and life began. The oldest rocks are over 800 million years old. There was little or no life on the Earth at that time. Simple plants and animals existed about 500 million years ago. Their remains have been found. The coal measures were formed over 200 million years ago. It is probable that there have been men on the Earth for about a million years, but this is a short period of time compared with the age of the Earth.

Fossils

The preserved remains of plants and animals found in certain rocks indicate the kinds of life that existed on the Earth in past ages. These remains are called *fossils.* Pieces of coal sometimes show impressions made by fossilized plants. Fossilized shells are sometimes found in chalk and limestone. Teeth and bones, millions of years old, are occasionally found in rocks and sand. Limestone, chalk and corals are animal skeletons. Oil is believed to be both animal and vegetable in origin. *Amber* is a fossilized resin.

Looking at fossils

Examine any available fossils and pieces of chalk, limestone, coral, coal, amber, etc. Which are plant and which are animal?

Extinct animals

Some of the giant reptiles which lived on the Earth about 200 million years ago are shown opposite. The *pterodactyl* was a flying reptile. These ferocious-looking creatures had small brains and were unable to compete successfully with the smaller but more intelligent mammals. Also, they were unable to adapt themselves to changes in climate, and so they became extinct.

How do we know what these animals looked like? Scientists have been able to learn about their sizes and appearance by reconstructing skeletons from fossilized bones.

Some kinds of animals have been made extinct in recent times by the activities of men. Large herds of *bison* once roamed the vast plains of North America. Today, there are only a few protected bison in Canadian parks. In Britain, certain birds are protected by Act of Parliament. The *dodo,* a flightless bird, last seen on the island of Mauritius, and the *quagga,* a zebra-like animal, became extinct quite recently. Some species of whales have been made extinct. There is the danger that whales and some other animals will become extinct unless the activities of hunters are more strictly controlled. As the human population becomes larger, human needs become greater; this is why it is important that we should give more attention to the conservation of our natural resources.

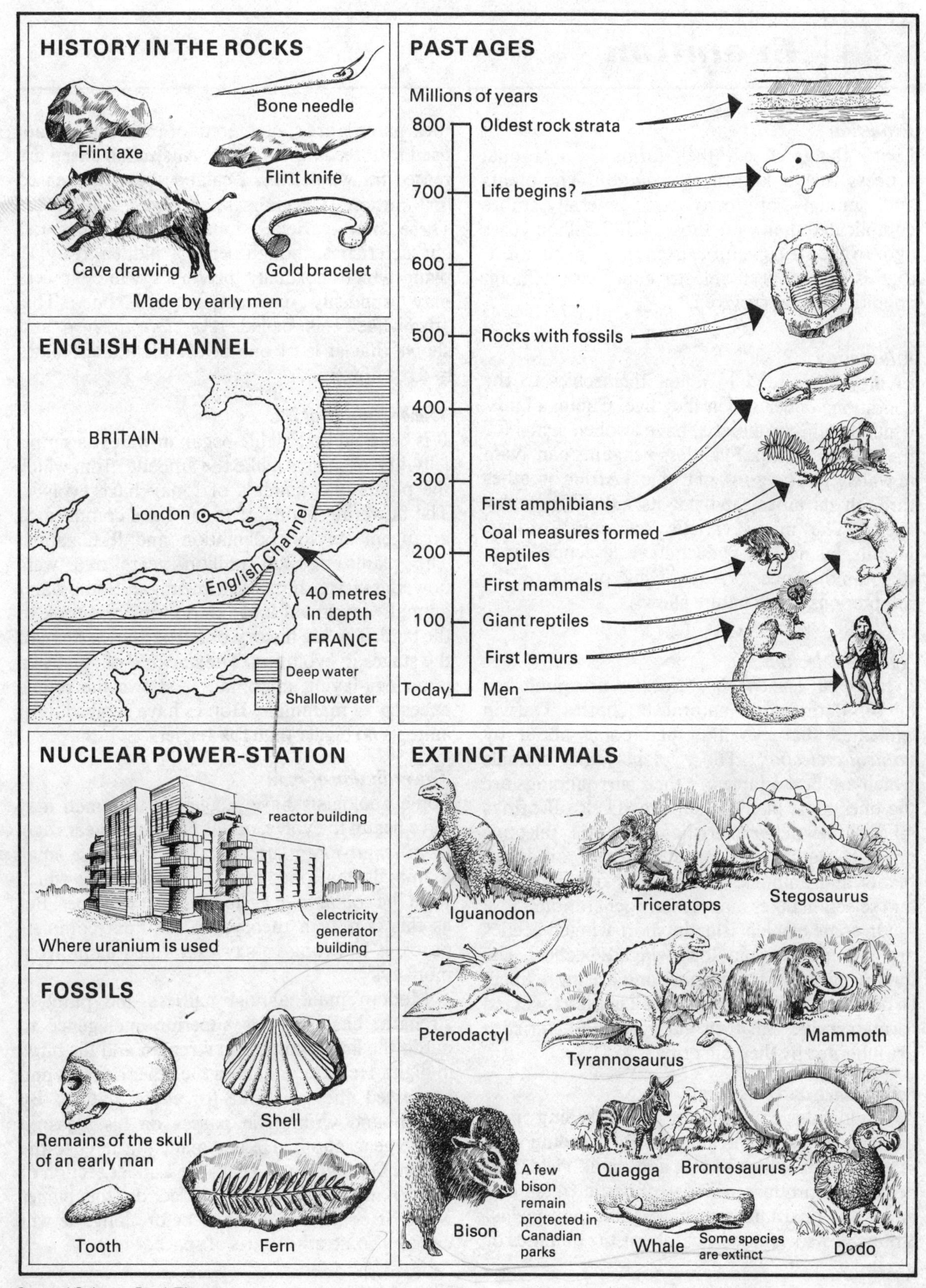
HISTORY IN THE ROCKS
Bone needle
Flint axe
Flint knife
Cave drawing
Gold bracelet
Made by early men
PAST AGES
Millions of years
800
Oldest rock strata
700
Life begins?
600
500
Rocks with fossils
400
300
First amphibians
Coal measures formed
200
Reptiles
First mammals
100
Giant reptiles
First lemurs
Today
Men
ENGLISH CHANNEL
BRITAIN
London
English Channel
40 metres
in depth
FRANCE
Deep water
Shallow water
NUCLEAR POWER-STATION
reactor building
electricity
generator
building
Where uranium is used
EXTINCT ANIMALS
Iguanodon
Triceratops
Stegosaurus
Pterodactyl
Tyrannosaurus
Mammoth
Bison
A few
bison
remain
protected in
Canadian
parks
Quagga
Brontosaurus
Whale
Some species
are extinct
Dodo
FOSSILS
Shell
Remains of the skull
of an early man
Tooth
Fern

30 Evolution

Evolution
Living things change their forms by a gradual process that is known as *evolution*. The plants and animals of today are generally more complicated than were those of 300 million years ago. In the struggle for existence, only the fittest, that is, those best able to cope with difficult conditions, have survived.

Adaptation
Living things tend to adapt themselves to the conditions under which they live. Cactus plants, which live in dry deserts, have swollen stems for the storage of sap. Flightless penguins can swim in water. The *lung-fish* of tropical Africa breathes through its lungs, and not its gills, during dry spells. The many varieties of dogs, wolves, jackals, hyenas, etc., which have descended from a common ancestor, are living proofs of the adaptations which nature allows.

Natural selection
In his book, *The Origin of Species*, first published in 1859, the great naturalist Charles Darwin explained that evolution had come about by *natural selection*. Those plants and animals which are best adapted to their surroundings are the ones most likely to survive. Their offsprings inherit their characteristics. By this selective process, nature has produced new kinds of plants and animals. The process of adaptation and selection takes many, many generations.

On some tropical islands, short-winged beetles are commoner than long-winged beetles. The long-winged forms fly higher and are more likely to be swept out to sea by winds. The short-winged forms survive and their survival characteristics are inherited by their offsprings.

Inheritance and breeding
Agriculturists copy nature by breeding new varieties of plants and animals. By breeding with selected cows which have a good milk yield, it is possible to produce varieties of cattle which can almost be guaranteed to have good milk yields. Young cows which do not inherit the qualities of their parents are slaughtered for meat and are not used for breeding purposes. Australian sheep are reared for wool; New Zealand sheep are reared for mutton. Racehorses are bred for speed. These new varieties inherit their ancestors' characteristics, though perhaps hidden away. A plant which normally produces white flowers may suddenly produce a red one. This "throwback" is called *reversion to type* and shows that at least one of the plant's ancestors had red flowers.

The beginning of life
It is believed that all life began in water as simple jelly-like organisms, like the amoeba, from which the plants and animals of today have evolved. The flowering plants of today have complicated arrangements for pollination and fertilization. The plants of 300 million years ago were flowerless ferns, horsetails and algae. Some water animals developed lungs and limbs and took to the land. The life history of a frog shows some of the stages in evolution. The *duckbill* of Australia is an egg-laying mammal. It shows the reptile ancestry of mammals. Horses have evolved from animals no bigger than fox-terriers.

The evolution of man
Some zoologists have suggested that men may have begun to evolve, about 60 million years ago, from *lemur-like* mammals, and that these small mammals developed into larger mammals which were the *common ancestors* of men and apes. But as this is only a theory, and without complete proof, it is wrong to say that "men came from monkeys".

Modern man, almost hairless and puny in strength, has used his superior intelligence to outdo the large animals in strength and the birds in flight. He has conquered the other animals and harnessed the powerful forces of nature. By speech and writing he passes on his ideas to future generations. He has made journeys to the Moon. On 20th July, 1976, the Lander section of the Viking I spacecraft landed on the planet Mars. It could well be that, before long, he will conquer some other parts of space.

ADAPTATION

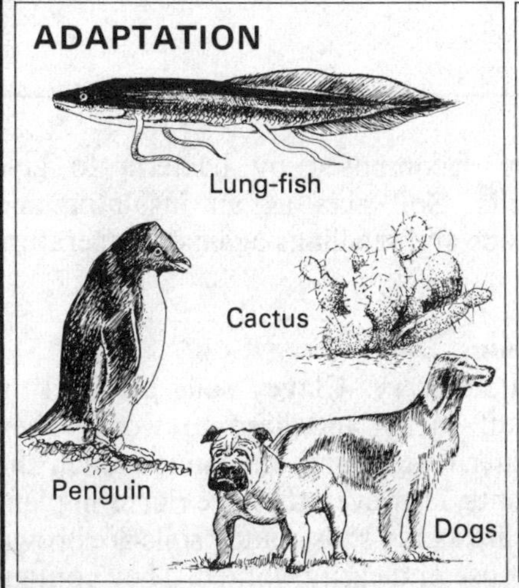

CHARLES DARWIN

1809-1882

Author, naturalist and explorer

INHERITANCE AND BREEDING

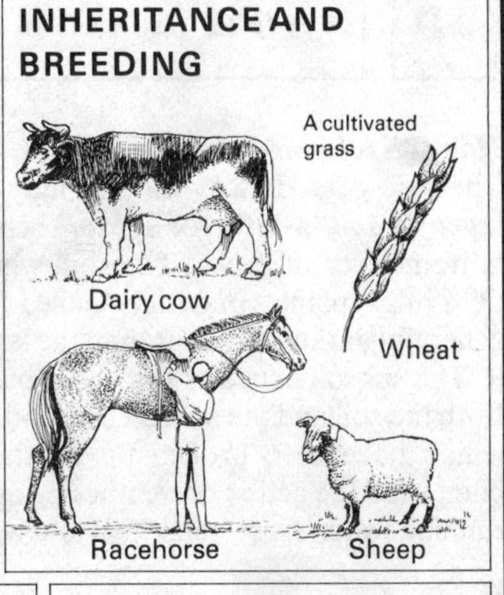

AMOEBA

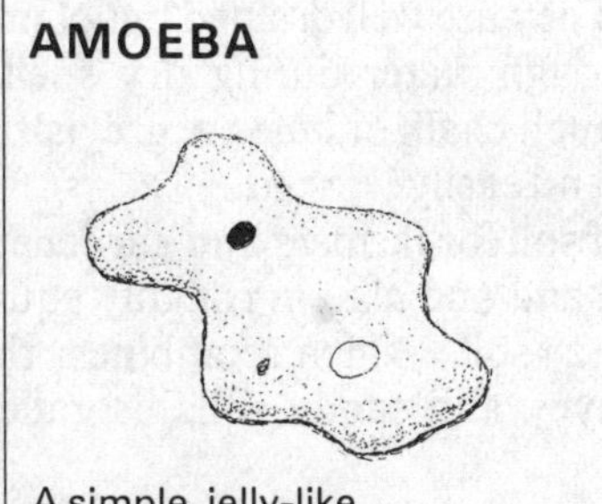

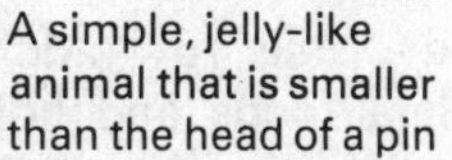

A simple, jelly-like animal that is smaller than the head of a pin

DUCK-BILLED PLATYPUS

An Australian mammal; sometimes called a *duck-mole* or *duckbill*

EARLY HORSE

Lived 60 million years ago and no bigger than a fox-terrier

THE DEVELOPMENT OF MAN

MODERN MAN

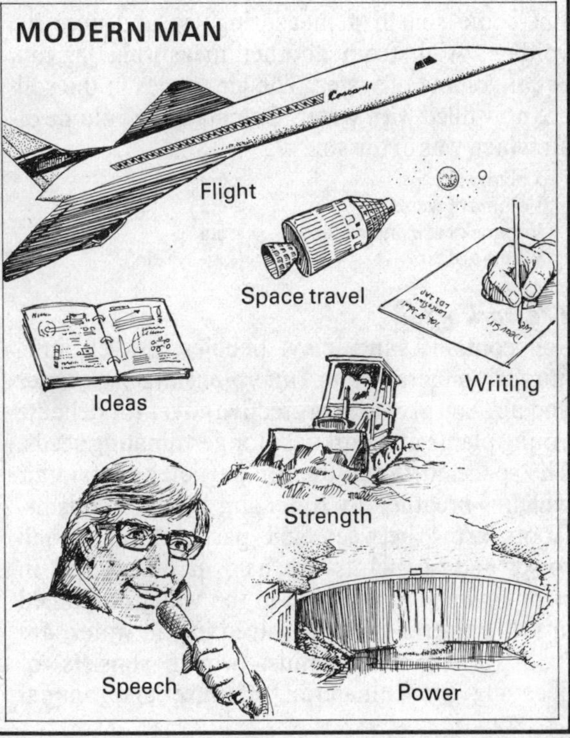

31 Soil

How soil is formed

The side of a railway cutting may show a top layer of *soil,* a layer of *subsoil* below it, and a bottom layer of rocks. The soil, which contains decaying plant material called *humus,* is generally darker in colour than the layers beneath it. The subsoil contains very little plant material. Both the soil and the subsoil are made from rocks which have been broken up by the growth of plants and the action of weathering agents—rain, running water, frost, solar heat and wind.

Soil contents

Shake up some soil in water in a jar. Allow it to settle. The heavier materials settle first. You can see separate layers of gravel, small pebbles, sand, clay and humus.

Soil water

Heat soil in a test-tube. Notice the "steam" which escapes from it. Test for water with blue cobalt chloride paper.

Soil air

Put some soil in a measuring jar and note its volume. Add, from another measuring jar, an equal volume of water. The air spaces in the soil are now filled with water. Calculate the volume of air which was in the soil.

Volume of soil	$= x\ cm^3$
Volume of water	$= x\ cm^3$
Volume of soil and water	$= y\ cm^3$
$\therefore$ Volume of air	$= (2x - y)\ cm^3$

The work of soil

Soil contains sand, clay, pebbles, gravel, lime, humus, mineral salts, *micro-organisms,* water and air. Soil provides protective cover for delicate young plants and darkness for germinating seeds. The spaces between the soil particles contain air which is breathed by roots and small organisms. Roots grow between soil particles and small rocks and so plants are held in place even in strong winds. Roots take in the water contained in soil. Mineral salts, dissolved in soil water, are food for plants. Humus, which consists of decaying plant material, manure and animal remains, is decomposed by bacteria to give mineral salts. Soil acts as an insulator and protects seeds and seedlings against temperature extremes.

Soil conditions

Soil conditions vary. Clayey soils are dark in colour, cold, sticky and hard to work. They become water-logged and air cannot reach the roots of plants. However, they are richer in plant foods than are sandy soils. Sandy soils are brown in colour, loose and easily worked. They contain many air spaces. They are well drained but plants do not obtain enough water during dry spells. Soils containing much chalk or lime are greyish in colour, dry, loose and easily worked.

The best kind of soil for farmers and gardeners is one containing sand and clay in roughly equal quantities. Such a soil, which combines the advantages of clayey and sandy soils, is called *loam.*

Farmers improve sandy soils by adding humus, which contains plant foods and holds moisture. Clayey soils are lightened by adding sand, ashes, lime and humus. Lime causes the clay particles to *flocculate,* that is, to stick together. These larger particles allow drainage and the soil is more easily worked.

Flocculence

Stir some sand in water in a beaker. The sand soon settles. Then stir some powdered clay in the water. The fine clay remains suspended. Now stir a little lime into the suspension. The clay particles flocculate and settle.

The work of earthworms

Charles Darwin realized the importance of earthworms in the soil. They make holes which assist in drainage and allow air to reach the roots of plants. They feed on vegetation, and the waste from their bodies, which is deposited on the ground surface as *casts,* is a valuable food for plants. The earthworms in a *hectare* (10 000 m^2) of cultivated ground lift between 20 000 kg and 50 000 kg of soil annually. No wonder that they are sometimes called "little ploughmen".

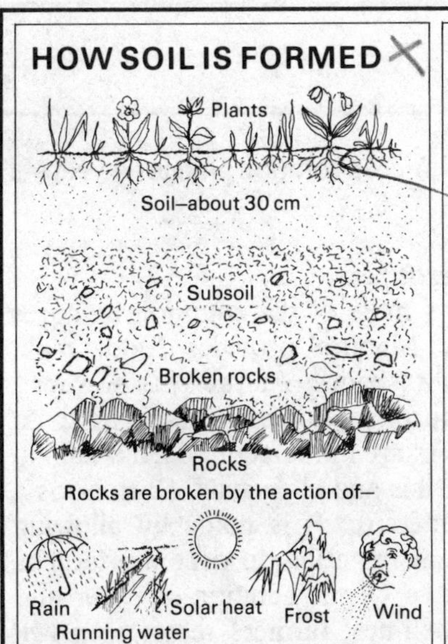
HOW SOIL IS FORMED
Plants
Soil–about 30 cm
Subsoil
Broken rocks
Rocks
Rocks are broken by the action of–
Rain
Running water
Solar heat
Frost
Wind

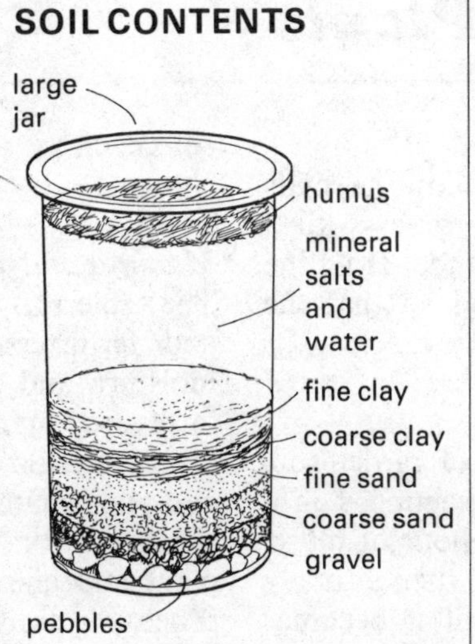
SOIL CONTENTS
large jar
humus
mineral salts and water
fine clay
coarse clay
fine sand
coarse sand
gravel
pebbles

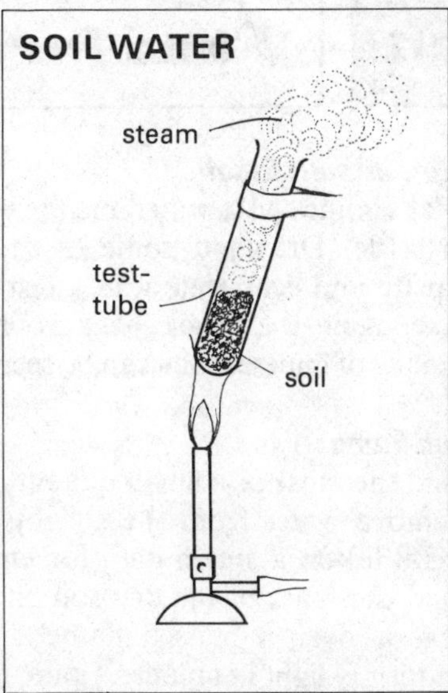
SOIL WATER
steam
test-tube
soil

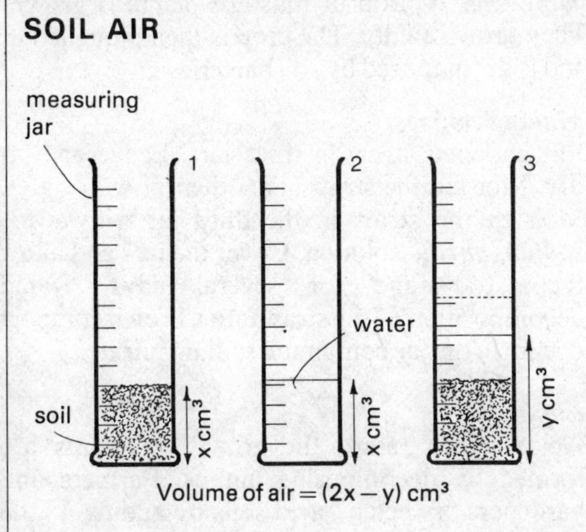
SOIL AIR
measuring jar
1
2
3
water
soil
x cm³
x cm³
y cm³
Volume of air = (2x – y) cm³

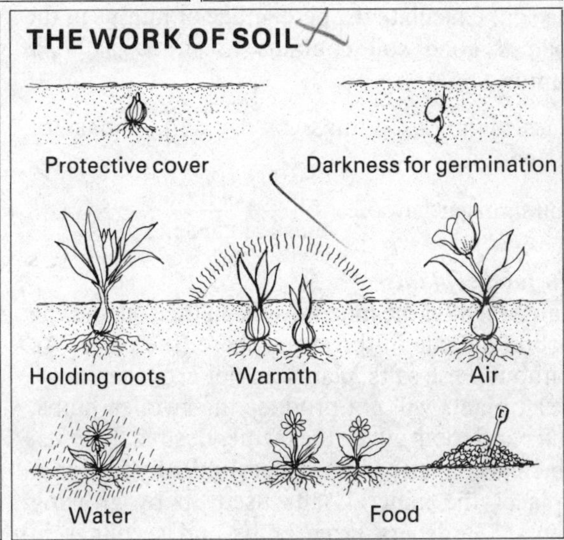
THE WORK OF SOIL
Protective cover
Darkness for germination
Holding roots
Warmth
Air
Water
Food

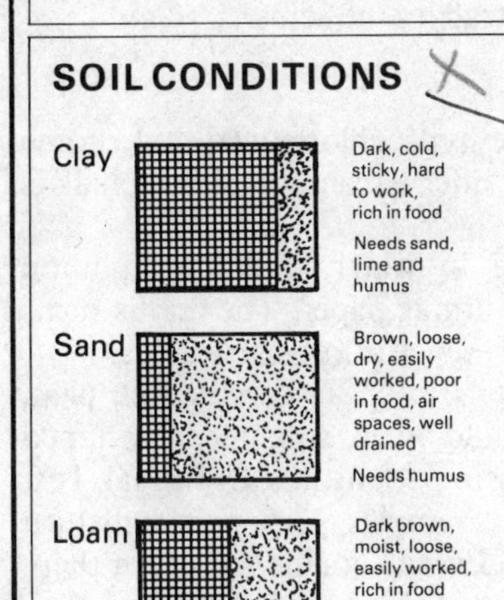
SOIL CONDITIONS
Clay
Dark, cold, sticky, hard to work, rich in food
Needs sand, lime and humus
Sand
Brown, loose, dry, easily worked, poor in food, air spaces, well drained
Needs humus
Loam
Dark brown, moist, loose, easily worked, rich in food
Best for farmers and gardeners

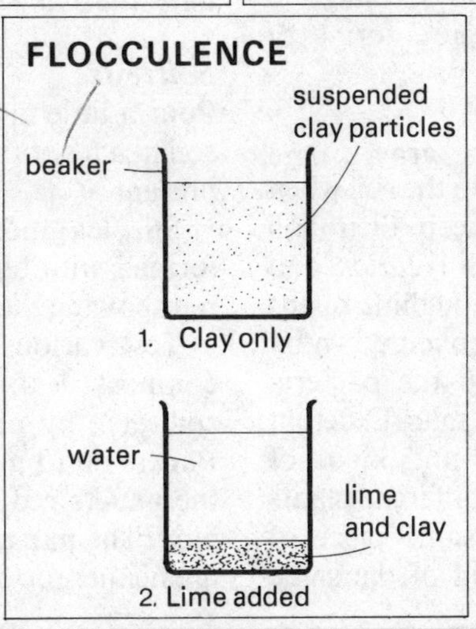
FLOCCULENCE
beaker
suspended clay particles
1. Clay only
water
lime and clay
2. Lime added

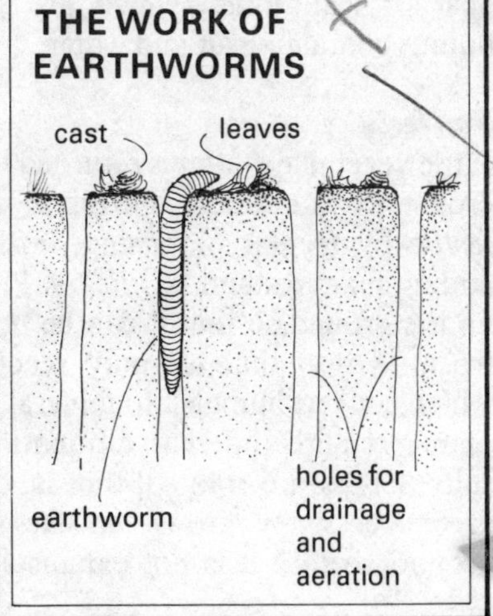
THE WORK OF EARTHWORMS
cast
leaves
earthworm
holes for drainage and aeration

32 Food for Plants

Mineral salts in soil

Shake some soil in water and then allow the solids to settle. Draw off some of the water with a pipette and evaporate it in a test-tube. Hold the tube against a black background so that the residue of mineral salts can be seen.

Soil humus

Find the mass of a small quantity of garden soil. Remove water from the soil by heating it in a metal lid on a steam bath for an hour or more. Find the mass of the dry soil and then strongly heat it over a bunsen flame until it becomes uniformly light in colour. Again, find the mass of the soil. Calculate the percentage of humus in the soil. A good soil contains as much as 10% humus.

Mass of humus = mass of dry soil − mass of heated soil

$$\text{Humus in garden soil} = \frac{\text{mass of humus}}{\text{mass of garden soil}} \times 100\%$$

The food of plants

Plants feed on the mineral salts in the soil and the carbon dioxide in the atmosphere (see page 74). Without these salts, plants do not grow healthily. Weak plants will not produce the swollen bulbs, fruits and roots which the farmer desires. Humus, which is converted into mineral salts by bacteria, replaces the mineral salts used up by growing plants. Gardeners grow bulbs and seedlings in peat or leaf-mould, which are pure forms of humus containing air and water.

Soil fertility

Little vegetable matter is returned to the soil when crops are harvested. Farmers keep their soil *fertile* by regular ploughing, *crop rotation* and adding manures and *fertilizers*. Ploughing opens up the ground so that it can be broken down by frost, and so that air may reach the bacteria which convert humus into mineral salts. Different plants require different amounts and kinds of salts. Crops are rotated, that is, different plants are successively grown on the same piece of ground so that it is not exhausted of the same salts each year. A crop rotation plan for 4 years is shown.

Manures and fertilizers

The table shows most of the common manures and fertilizers. Manure provides food, bulk, moisture and air spaces. At present, however, farmyard manure is expensive and hard to obtain. Fertilizers are manufactured from animal and plant materials and chemicals. *Compost* is a substitute for manure. It is made by allowing plant and animal materials to decompose. The process is assisted by the addition of water and chemicals. Sometimes, farmers resort to *green manuring*. A crop of mustard plants is grown. They grow rapidly. The crop is then ploughed in and is decomposed by soil bacteria.

Making compost

Put chopped straw in three jars. Leave one jar dry. Moisten the straw in another jar with water. Moisten the straw in the third jar with a 1% *sodium nitrate* solution. Cover the jars and allow them to stand for several days. Rapid decomposition of the straw into a brown compost occurs in the jar containing sodium nitrate.

Liming

Acids which retard the growth of plants are formed by decomposing humus. Farmers and gardeners "sweeten" acid soils by adding a little lime; the lime neutralizes the acids.

Soil tests

Pour a little dilute hydrochloric acid on to some soil in a test-tube. Effervescence occurs if chalk is present.

Stir leaf-mould in water. Test the solution formed with blue litmus paper. The litmus turns red showing the presence of vegetable acids.

Test various soils, such as garden soil, peat, compost, leaf-mould, field soil, recently limed soil, etc., by pressing Johnson's Universal Test Papers on to moist samples. Very acid soils turn the papers red and very alkaline soils turn them blue. The papers are turned green by soils which are neither too acid nor too alkaline.

MINERAL SALTS IN SOIL

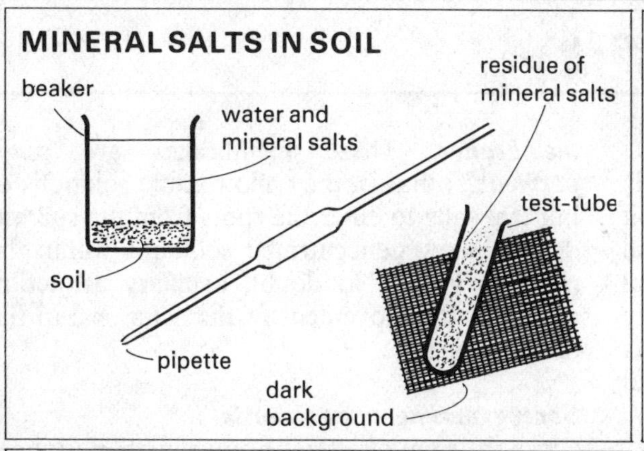

SOIL HUMUS

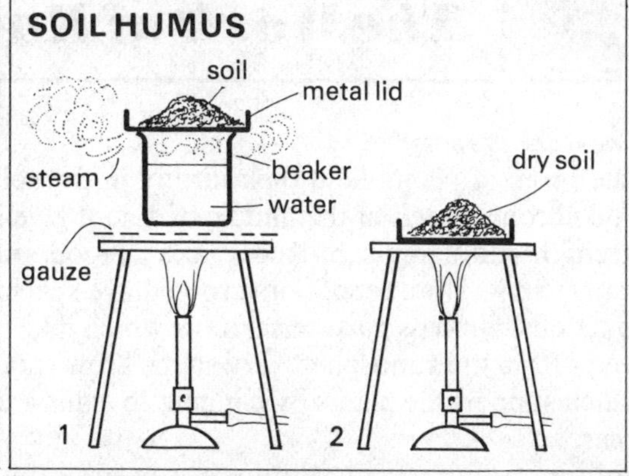

MANURES AND FERTILIZERS

Vegetable		Natural humus	Decaying leaves, stems, etc.
		Compost	Decaying plants, produced artificially
		Animal manure	From pigs, horses, sheep, cows, poultry, etc.
		Seaweed	Used near the coast
		Hops	By-product of brewing
		Green manure	A crop of mustard or rape
		Soot	Absorbs heat (see page 54, Book Two)
		Bonfire ash	Rich in potash
Animal		Dried blood	From slaughter-houses
		Ground hoofs	
		Ground bones	Take some time to decompose
		Ground horns	
		Shoddy	Woollen waste
		Leather and hair	By-products of tanning
		Guano	Bird droppings and feathers (from Peru and Chile)
		Fish meal	Fish that is unfit for food
Chemical		Basic slag	
		Sulphate of ammonia	By-products of iron smelting
		Superphosphate	
		Sodium nitrate (nitrate of soda or Chile saltpetre)	Guano is mainly nitrates
		Lime	Sweetens acid soils

FOUR-YEAR CROP-ROTATION PLAN

Root vegetables	Oats	1976
Wheat	Clover	
Wheat	Root vegetables	1977
Clover	Oats	
Clover	Wheat	1978
Oats	Root vegetables	
Oats	Clover	1979
Root vegetables	Wheat	

A different crop is grown in each field for 4 years

MAKING COMPOST

SOIL TESTS

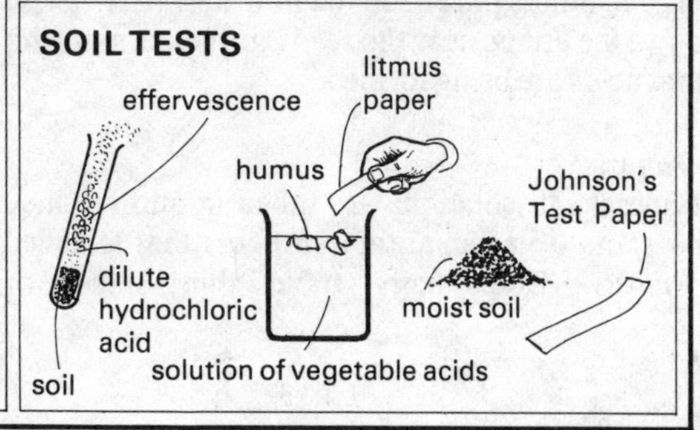

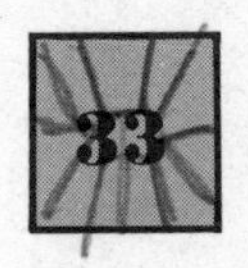

The Work of Roots

The work of roots
The roots of plants hold them firmly in the soil. The fibrous *vessels* in the middle of a root give it strength and carry sap. Roots take in food and water. Some store food. Some roots have special functions. Orchids have aerial roots which take in water from the atmosphere. Ivy stems grow roots which support the plants by clinging to walls and trees.

Root vessels
Stand a dandelion plant in red dye. After a few hours, cut off a portion of the root and examine it under a magnifying glass. The vessels in the centre of the root are filled with red dye.

Root growth
Roots grow downwards (geotropism) and towards water (hydrotropism). The growth of a root occurs behind its delicate tip, which is protected by a *root cap*. Behind the cap, there are *root hairs*. They take in water and mineral salts. Root hairs cling to soil particles and absorb the thin films of moisture around them. When transplanting, it is important to transfer the soil around the roots so that the root hairs are not damaged. Growing roots produce weak acids.

Some experiments on root growth
Grow a few broad beans on moist cotton wool in a jar. Examine one of the root tips under a hand-lens. Make a labelled drawing.

Use scissors to remove a root tip. Return the seedling to the jar. Examine it after a few days. Has it continued to grow?

Germinate radish or mustard seeds on pieces of blue litmus paper soaked in water. After a few days, the litmus near the seeds turns red, showing that acids are being formed.

Osmosis
Mineral salt solutions are taken in and retained by roots by a remarkable process that is called *osmosis*. Root hairs have thin skins, or *membranes*. These membranes are *semi-permeable,* that is, they allow dilute solutions of mineral salts to enter the roots from the soil but do not allow concentrated solutions within the roots to leave. No doubt, capillary attraction assists in the movement of the solutions up the roots and stems.

Some experiments on osmosis
Close the stem of a thistle-funnel with plasticine. Half fill the funnel and its stem with sugar solution and cover it with parchment or cellophane held in place by a rubber band. Invert the funnel and remove the plasticine. Immerse the funnel in water and allow it to stand for a few hours. The solution rises up the stem. The water passes through the semi-permeable membrane but the solution does not.

Place a slice of potato in a strong salt solution. After a few hours, it goes limp. Water leaves the slice through the potato membranes but the salt solution cannot enter it. Place the limp slice in water. The reverse process now happens; the slice swells as water enters it.

Immerse an egg in dilute hydrochloric acid to dissolve away the shell. Place the egg in a salt solution. It decreases in size. Now place the egg in water. It swells up beyond its normal size. Why?

Place the roots of a plant in a strong salt solution. The plant wilts. Why?

Soil erosion
Roots help to hold soil together. The wholesale destruction of forest trees and their roots, constant cropping and inadequate manuring have resulted in the formation of dry, light, infertile soils which are easily blown away by winds and washed away by water. *Soil erosion* is prevented by *terracing, strip-cropping* and planting trees, bushes and certain types of grasses with many fibrous roots. In strip-cropping, strips of land are left uncropped; these strips act as barriers and prevent loose soil from being blown away by winds and washed away by water.

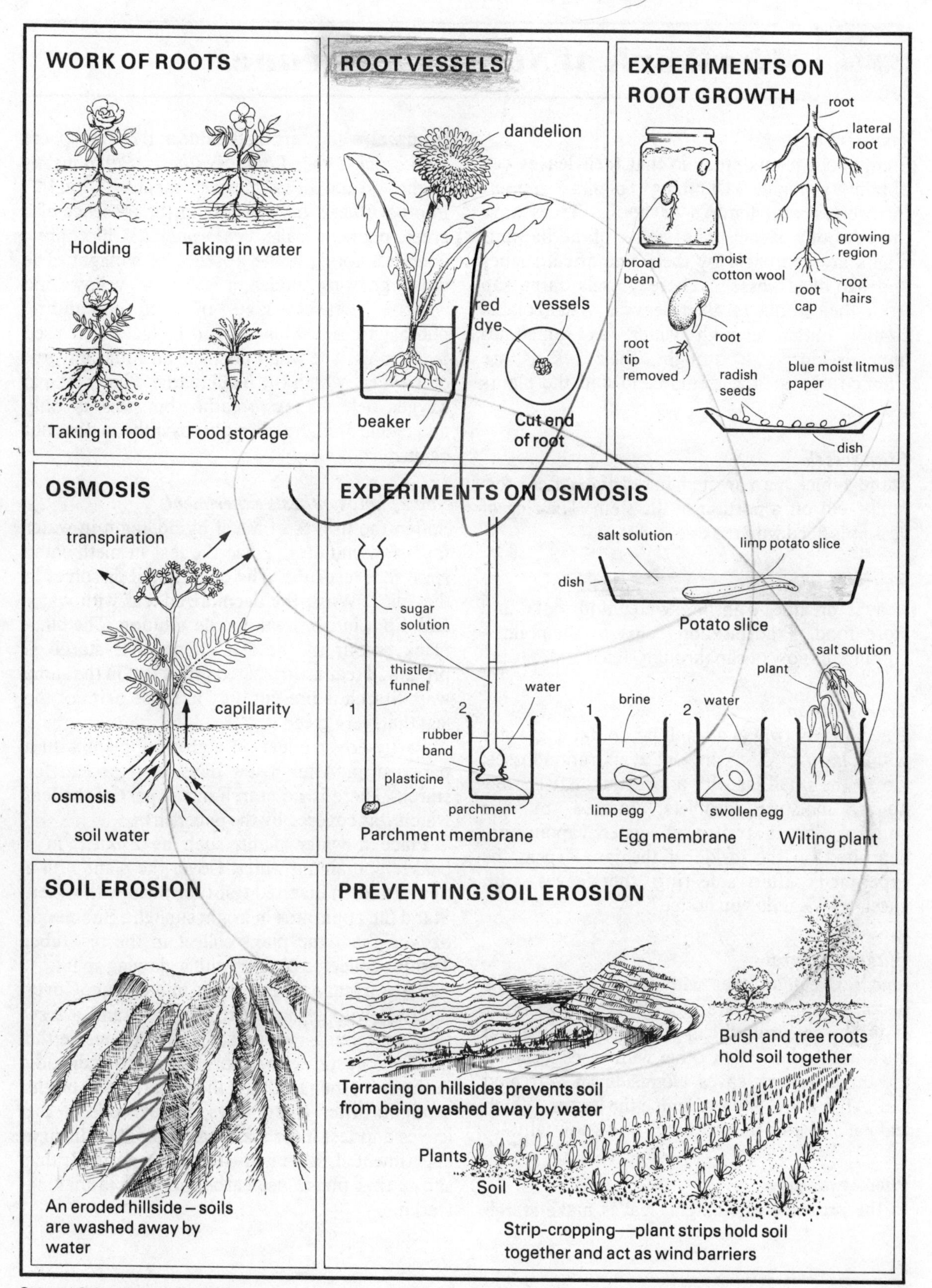
WORK OF ROOTS
Holding
Taking in water
Taking in food
Food storage
ROOT VESSELS
dandelion
red dye
vessels
beaker
Cut end of root
EXPERIMENTS ON ROOT GROWTH
root
lateral root
growing region
root cap
root hairs
broad bean
moist cotton wool
root
root tip removed
radish seeds
blue moist litmus paper
dish
OSMOSIS
transpiration
capillarity
osmosis
soil water
EXPERIMENTS ON OSMOSIS
sugar solution
thistle-funnel
1
2
water
rubber band
plasticine
parchment
Parchment membrane
salt solution
limp potato slice
dish
Potato slice
brine
water
limp egg
swollen egg
Egg membrane
plant
salt solution
Wilting plant
SOIL EROSION
An eroded hillside – soils are washed away by water
PREVENTING SOIL EROSION
Terracing on hillsides prevents soil from being washed away by water
Bush and tree roots hold soil together
Plants
Soil
Strip-cropping—plant strips hold soil together and act as wind barriers

34 The Work of Stems and Leaves

The work of stems

Stems hold plants erect so that their leaves can obtain sunlight. Climbing plants support themselves with tendrils or hooks. Trees have thick woody stems. The stems of herbaceous plants are kept rigid by their cylindrical shapes and sap-filled vessels. These vessels carry sap from their roots to their leaves. When plants cannot obtain enough water, as sometimes happens during a drought, their vessels are emptied of sap; the stems bend and the plants wilt.

Stem vessels

Stand a thick marrow stem in red dye. After a few hours, cut off a portion of the stem. The ring of vessels is filled with red dye.

The work of leaves

Leaves breathe, transpire water and make and store food. Transpiration helps to maintain a continuous flow of sap through stems and leaves.

Transpiration

Place a leafy twig in a small jar containing water with a layer of oil to prevent evaporation losses. Cover the small jar with a large jar. After a few hours, moisture droplets, formed by the condensation of transpired water vapour, are deposited on the inside of the jar. Repeat the experiment after smearing the leaves with vaseline. What do you notice?

Withering leaves

Smear a leaf all over with vaseline so that its pores, or *stomata,* are closed. Treat two other leaves by smearing the upper surface of one and the lower surface of the other with vaseline. Hang the three treated leaves alongside an untreated leaf. After a few days, examine the leaves. Which leaf has withered the most? Do you know why?

Photosynthesis

In the presence of sunlight, leaves make starch from water and carbon dioxide in the atmosphere by a process called *photosynthesis*. *Photo* means "light" and *synthesis* means "build". Photosynthesis only occurs in the presence of a green pigment called *chlorophyll.* Chlorophyll does not form in the absence of sunlight. The grass growing under a stone is yellow and weakly. Leaves give off oxygen during photosynthesis. This oxygen replaces that used by animals. The water plants in an aquarium give off oxygen, which the fish breathe. Plants take in oxygen when they breathe, but this is only noticeable at night when photosynthesis does not occur.

Some photosynthesis experiments

Soften the tissues of a leaf by boiling it in water for a few minutes. Place the leaf in methylated spirit in a test-tube. The chlorophyll dissolves in the spirit. Wash the decolorized leaf with water and dip it into a weak iodine solution. The blue-black colour of the leaf shows that starch is present. Treat a variegated privet leaf in the same way. Starch is present only in those parts of the leaf that were green.

Partly cover a leaf on a growing plant with a paper strip. After a few days, test the leaf for starch. There is no starch in the part of the leaf which was covered by the paper strip.

Place a water plant, such as *Elodea,* in a beaker containing water. Cover the plant with a funnel and an inverted test-tube filled with water. Stand the apparatus in bright sunlight. Bubbles of oxygen from the plant collect in the test-tube. Test for collected oxygen with a glowing splint.

Place a leafy twig in water in a trough. Cover the twig with an inverted gas-jar. Stand a test-tube containing caustic soda solution inside the jar. The caustic soda absorbs the carbon dioxide in the air in the jar. Stand the apparatus in bright sunlight. After a few days, remove one of the leaves and test it for starch as you did in the first experiment. Little or no starch has formed; this shows that plants use carbon dioxide in making starch.

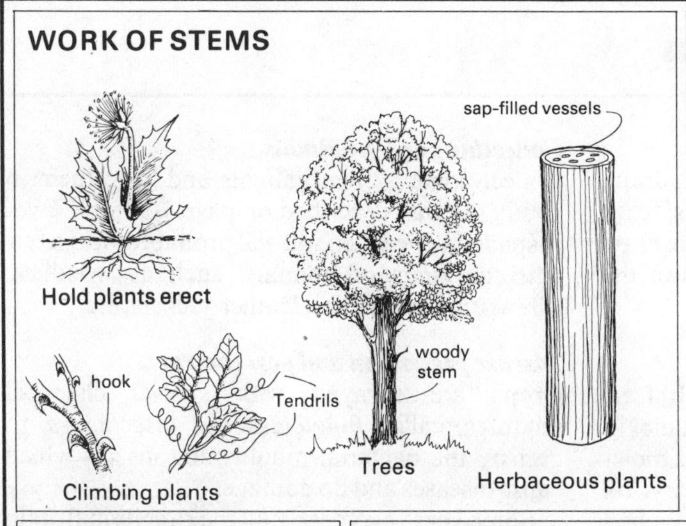
WORK OF STEMS
sap-filled vessels
Hold plants erect
hook
Tendrils
woody stem
Trees
Herbaceous plants
Climbing plants

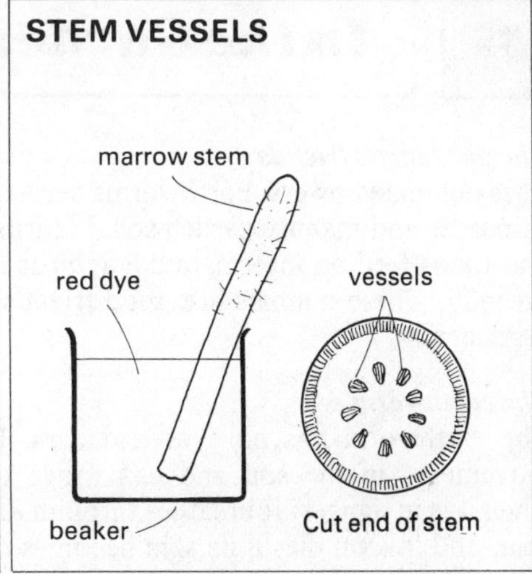
STEM VESSELS
marrow stem
red dye
vessels
beaker
Cut end of stem

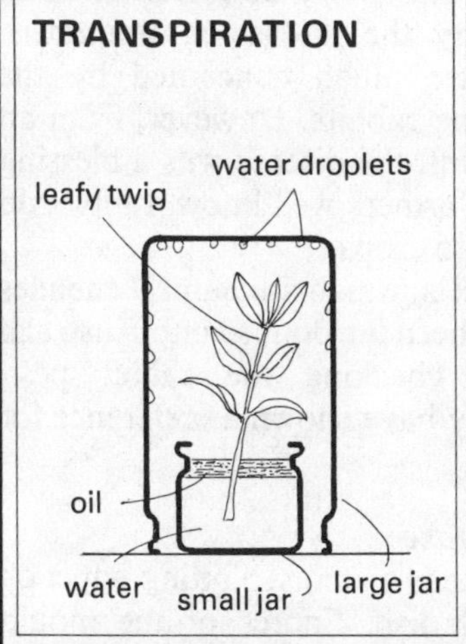
TRANSPIRATION
water droplets
leafy twig
oil
water
small jar
large jar

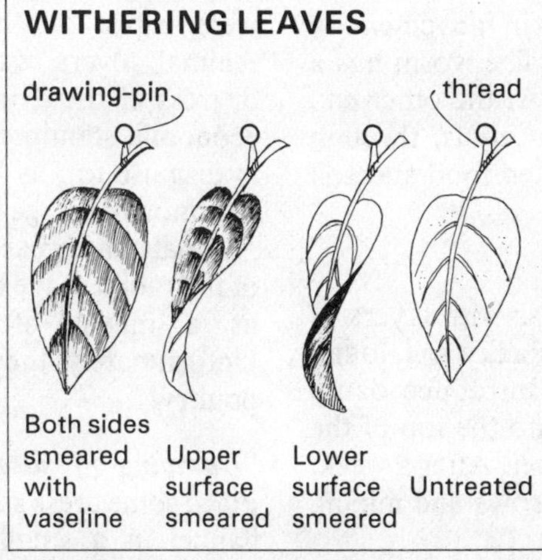
WITHERING LEAVES
drawing-pin
thread
Both sides smeared with vaseline
Upper surface smeared
Lower surface smeared
Untreated

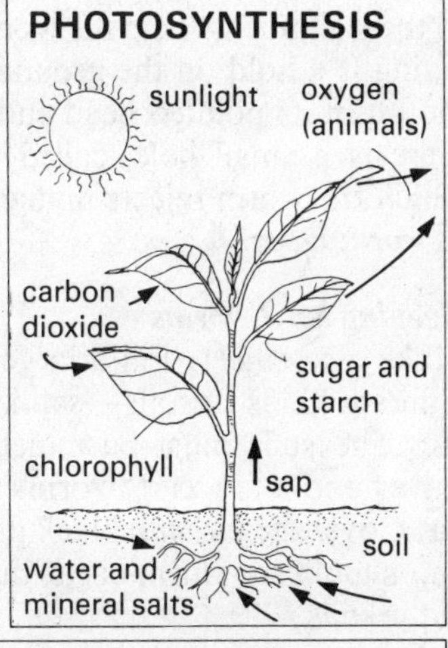
PHOTOSYNTHESIS
sunlight
oxygen (animals)
carbon dioxide
sugar and starch
chlorophyll
sap
soil
water and mineral salts

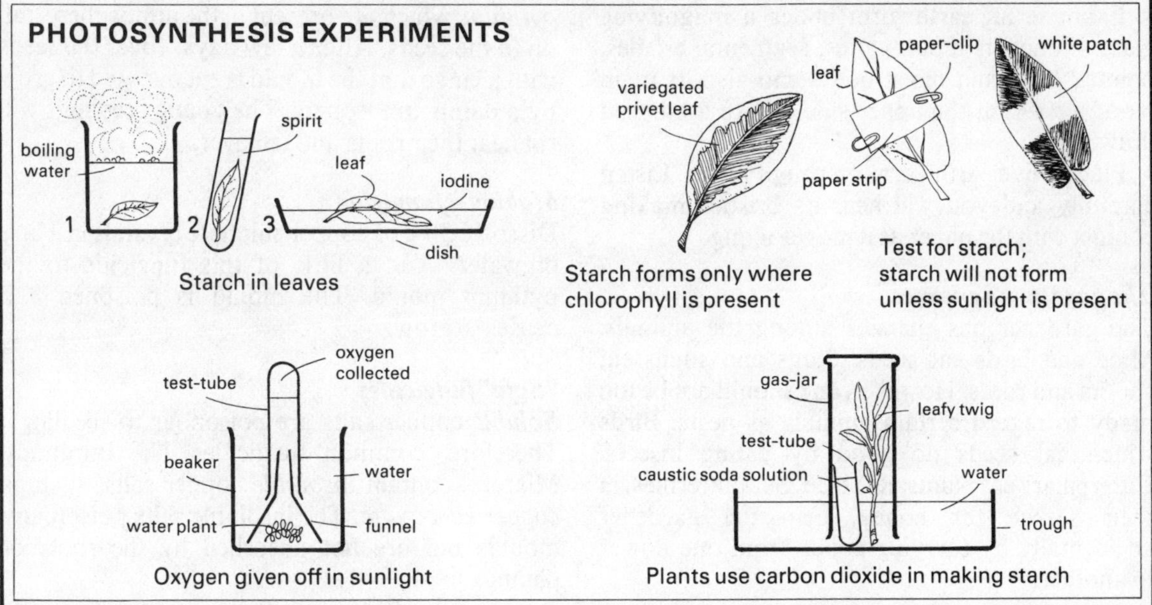
PHOTOSYNTHESIS EXPERIMENTS
boiling water
spirit
leaf
iodine
dish
1
2
3
Starch in leaves
variegated privet leaf
Starch forms only where chlorophyll is present
leaf
paper-clip
white patch
paper strip
Test for starch; starch will not form unless sunlight is present
oxygen collected
test-tube
beaker
water
water plant
funnel
Oxygen given off in sunlight
gas-jar
leafy twig
test-tube
caustic soda solution
water
trough
Plants use carbon dioxide in making starch

35 In the Garden

The gardener's friends

Bees pollinate flowers. Earthworms aerate, drain, turn over and manure garden soil. Lizards, frogs and toads feed on insects, and ladybirds feed on greenfly. These animals are good friends of the gardener.

The earthworm

The earthworm has no eyes and ears. It feels movements in the soil and can move quickly when it is in danger. It breathes through its moist skin, and it soon dies if its skin becomes dry. Its body is divided into *segments*. *Bristles* attached to these segments help the worm in movement by giving it a hold on the ground. The worm has a mouth in its pointed head end. At the other end there is a small hole, called the *anus*, through which the worm rejects undigested food and soil as *worm-casts*.

Keeping earthworms

Make a *wormery*. Fill a large jar with layers of different kinds of soils—sand, chalky soil, loam, etc. The soils must be moist. Introduce damp leaves and a few earthworms into the top of the jar. Cover the jar with black paper. After a week, you should see worm-casts, burrows and mixing of the soils.

Examine an earthworm under a magnifying glass. Notice its mouth, anus, segments, bristles, pointed head and moist body, and also its main blood vessel on the upper side. Make a labelled drawing.

Place a live earthworm on rough paper. Listen carefully and you will hear its bristles making contact with the paper as it moves along.

The gardener's enemies

The gardener has enemies among the animals. Mice and birds eat seeds. Slugs and snails eat shoots and roots. However, one should not be too ready to regard certain animals as pests. Birds which eat seeds do good by eating insects. Caterpillars eat plants, it is true, but butterflies, in their search for honey, help the gardener accidentally by carrying pollen from one flower to another.

Collecting garden animals

Collect some garden animals and keep them in jars in the way described on page 70, Book Two. A spadeful of garden soil will probably yield a few different kinds of animals, such as woodlice, wireworms, millipedes, leather-jackets, etc.

Disease prevention and pest control

Crops are sprayed with special chemical mixtures, called *fungicides* and *insecticides*, to destroy the bacteria, moulds and insects which cause diseases and do damage.

Some years ago nearly all the rabbits in Britain were wiped out by the disease *myxomatosis*. Animal lovers were much concerned by the distress of the dying rabbits. However, from an economic standpoint, the disease was a blessing in disguise for, as farmers well know, rabbits do enormous damage to crops.

Stoats and weasels, which are natural enemies of the rabbit, have been introduced into Australia as a means of checking the rabbit pest. Unfortunately, they have shown a preference for poultry!

"Damping-off" disease

Place some cress seeds on moist blotting-paper or flannel in a small dish. Spores of the mould *pythium*, which are present in the atmosphere, fall on to the seeds. After a few days, cover the seeds with a jar so that the mould is encouraged to grow by a damp atmosphere. The young seedlings will rot near their roots and fall over.

Making a fungicide

Dissolve 30 g of copper sulphate crystals in a litre of water. Add a little of this fungicide to the pythium mould. The mould is poisoned and ceases to grow.

"Safe" fungicides

Soluble copper salts are poisonous to seedlings. Therefore, common fungicides, like Burgundy Mixture, contain *insoluble* copper salts, such as *copper carbonate*. The insoluble salts poison any moulds but are not absorbed by the roots of plants.

THE GARDENER'S FRIENDS

EARTHWORM

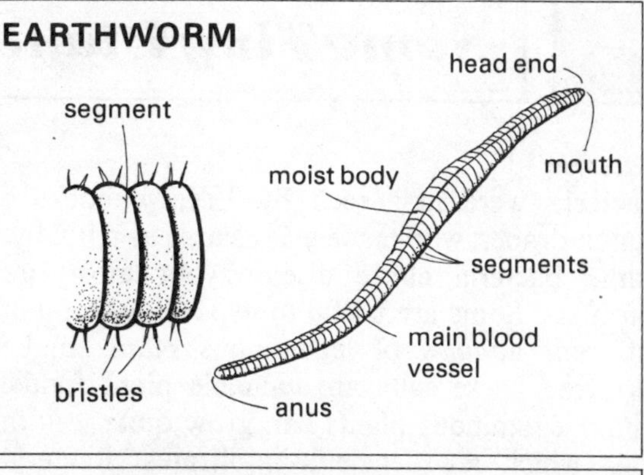

KEEPING EARTHWORMS

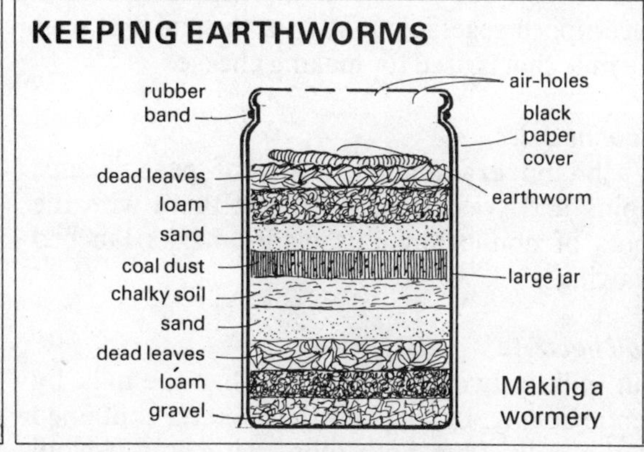

THE GARDENER'S ENEMIES

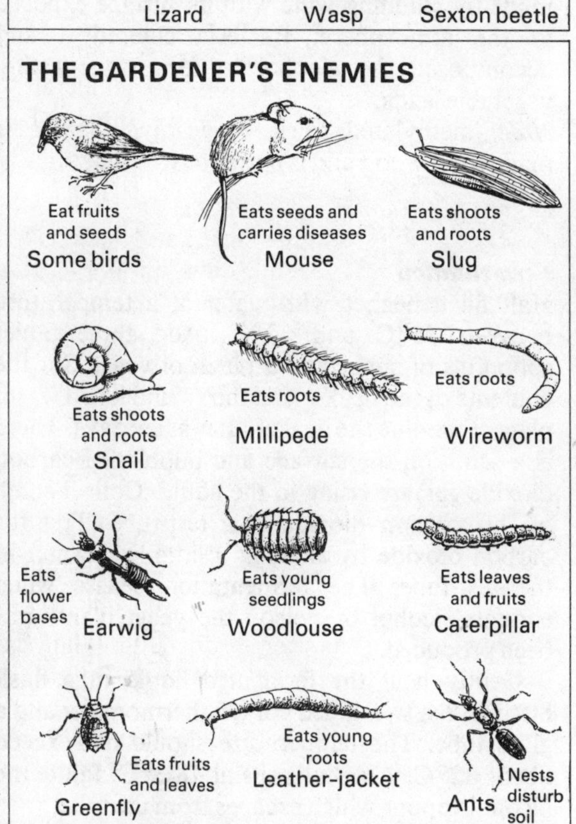

DISEASE AND PEST CONTROL

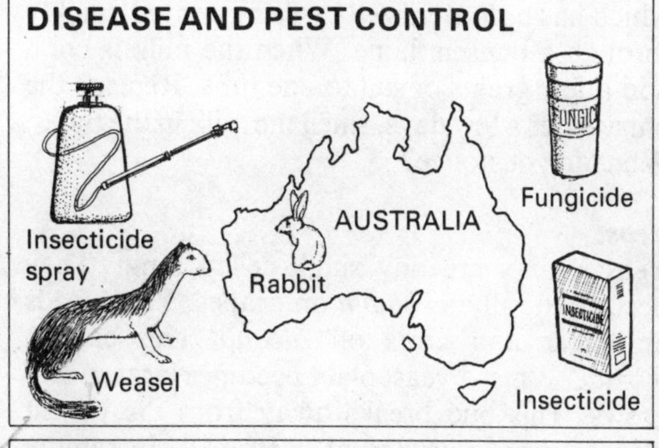

"DAMPING-OFF" DISEASE

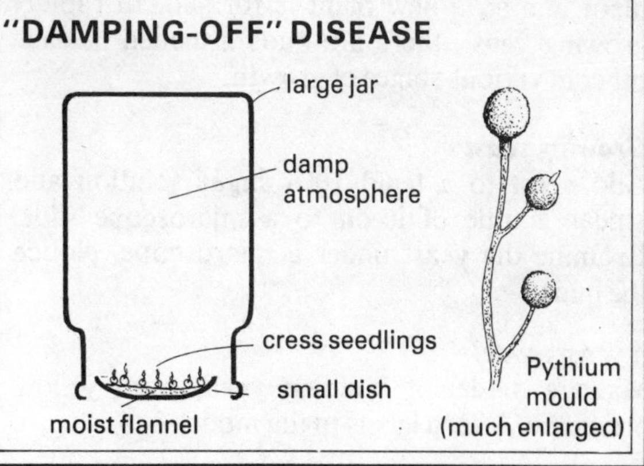

36 Some Tiny Plants

Bacteria
Bacteria were first seen by Leeuwenhoek, a Dutch draper, who made glass lenses as a hobby. Some bacteria cause diseases but most are harmless. Some are useful to us. The bacteria in the *root nodules* of leguminous plants make *nitrates*. These salts are valuable plant foods. Thus, leguminous plants can grow quite well in soils which are deficient in nitrates. Bacteria decompose vegetable matter in the soil and sour the milk that is used for making cheese.

Root nodules
Wash and examine the roots of peas, beans, lupins and sweet peas. Contrast them with the roots of non-leguminous plants. Make labelled drawings.

Soil bacteria
Put milk in two test-tubes. Sterilize the milk by gentle boiling. Boiling kills the bacteria contained in the milk. Plug both tubes with cotton wool which has been sterilized by being passed quickly through a bunsen flame. When the milk is cool, add a few grains of soil to one tube. Replace the plug. After a few days, smell the milk in the tubes. What do you notice?

Yeast
Yeast plants are tiny single-celled fungi. They occur naturally as *bloom* on grapes. Yeast feeds on sugar and gives off *alcohol* and carbon dioxide. When a yeast plant becomes large, a *bud* grows. This bud breaks away from the parent plant and so a new plant is formed. In rapidly growing yeast, there are buds attached to each other in various stages of growth.

Growing yeast
Add yeast to a tepid 10% sugar solution and smear a little of it on to a microscope slide. Examine the yeast under a microscope. Notice the buds.

A yeast model
Make a model of budding yeast from yellow plasticine. Attach labels to the model.

Bread and alcohol
Bakers add yeast to the dough for making bread. The yeast feeds on the sugars in the dough and produces carbon dioxide, which causes it to rise and acquire a spongy texture.

Fruit, sugar, yeast and water are used in making wine. The yeast feeds on the sugar and gives alcohol. It is the alcohol in wine which causes *intoxication*. If wine is bottled before *fermentation* has ceased completely, carbon dioxide gas dissolves in the wine, which sparkles and fizzes when the bottles are opened. Fruit juices give wines their distinctive flavours.

Spirit, which is almost pure alcohol, is made by distilling wine. Brandy is distilled from wine. Gin is distilled from beer. Beer is made from yeast, hops and barley. Germinating barley seeds, which are called *malt*, produce sugar. Alcohol has many industrial uses as a solvent. It is used as a stimulant in medicine. At one time, vinegar was made by standing wine with its surface exposed to the atmosphere. Bacteria entered it and decomposed the alcohol to form sour-tasting vegetable acids.

Note: methylated spirit, or *methyl alcohol*, is made from wood and is *poisonous*.

Fermentation
Half fill a beaker with water at a temperature between 15°C and 33°C. Add three table-spoonfuls of sugar and a pinch of yeast. Stir the contents of the beaker and then stand it in a warm place. Examine the beaker after a few days. There is a scum on the surface and bubbles of carbon dioxide gas are rising in the liquid. Collect some of this carbon dioxide in a test-tube. Test for carbon dioxide by shaking a little lime-water in the test-tube. The fermentation ceases when enough alcohol to poison the yeast plants has been produced.

Gently heat the fermented liquid in a flask fitted with a two-holed cork, a thermometer and a glass tube. The temperature should not exceed about 82°C. Alcohol boils at 78.3°C. Ignite the alcohol vapour which escapes from the tube.

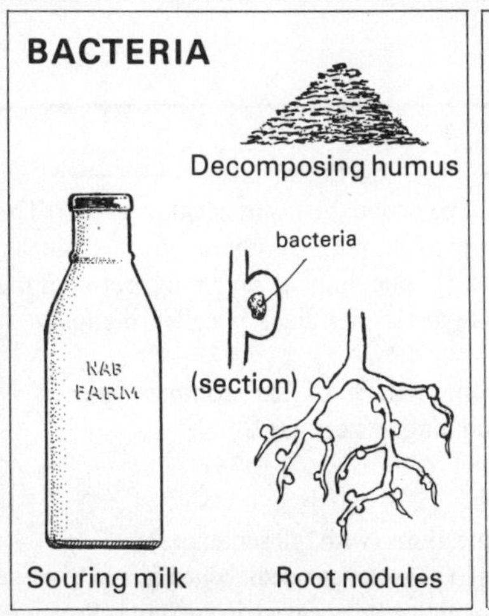
BACTERIA
Decomposing humus
bacteria
NAB FARM
(section)
Souring milk
Root nodules

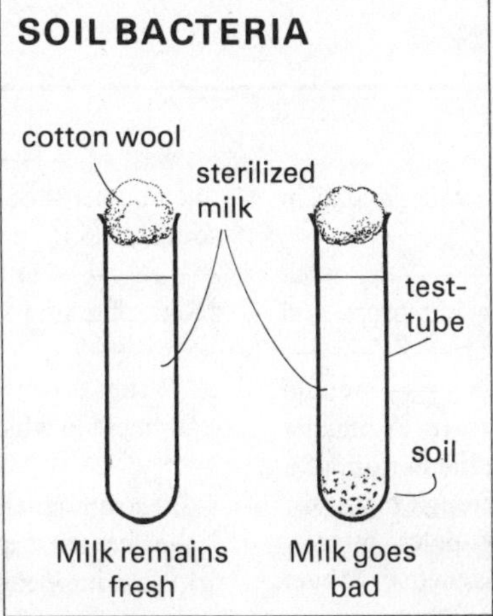
SOIL BACTERIA
cotton wool
sterilized milk
test-tube
soil
Milk remains fresh
Milk goes bad

ANTONY VAN LEEUWENHOEK
1632-1723
He was the first man to see bacteria

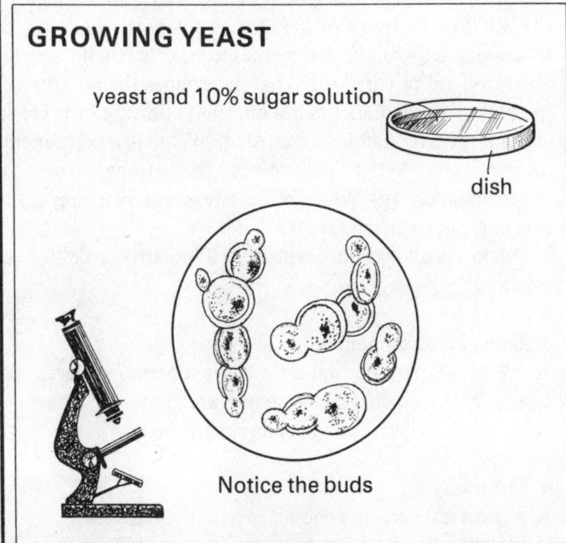
GROWING YEAST
yeast and 10% sugar solution
dish
Notice the buds

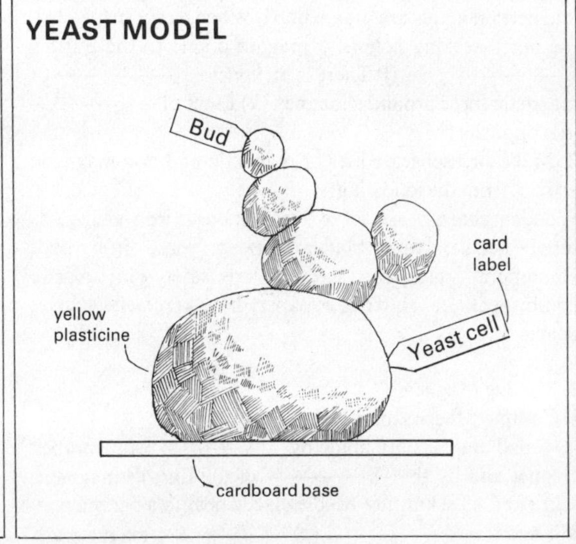
YEAST MODEL
Bud
card label
yellow plasticine
Yeast cell
cardboard base

BREAD AND ALCOHOL
Bread
Beer
Wine
carbon dioxide
sugar
alcohol
Brandy
Gin
Vinegar
Yeast

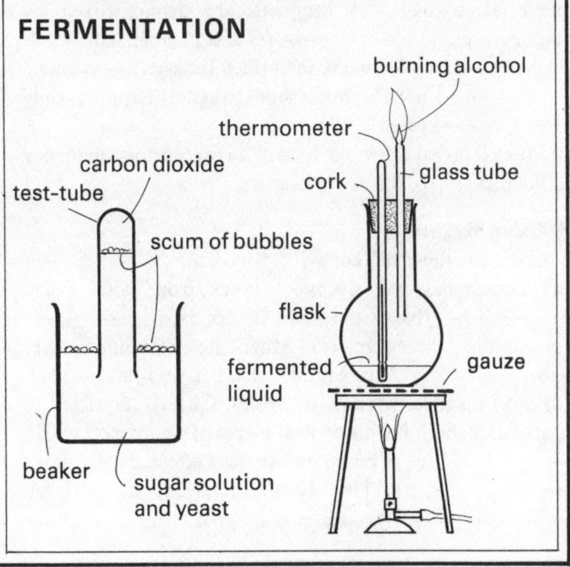
FERMENTATION
burning alcohol
thermometer
glass tube
cork
carbon dioxide
test-tube
scum of bubbles
flask
fermented liquid
gauze
beaker
sugar solution and yeast

Exercises

1 Magnets

1. Complete these sentences. (Copy the sentences and fill in the blank spaces with suitable words.)
(a) Magnets attract objects made of ———— or steel. (b) Steel and ———— are magnetic but brass and ———— are not. (c) A suspended bar ———— points in a north-south direction. (d) A ———— could be used to separate steel and brass screws. (e) A magnet will pick up an iron nail which is lying on the bottom of a dish of water because ———— acts through materials. (f) Like poles repel and ———— poles attract. (g) ———— magnets retain their magnetism longer than bar magnets. (h) Iron, cobalt, nickel, ———— and certain alloys are magnetic. (i) When freely suspended, the north-seeking pole of a magnet points to the Earth's ———— pole. (j) There is an invisible ———— of magnetic force around a magnet. (k) Lines of ———— join opposite poles.
2. Make *two* separate lists of magnetic and non-magnetic articles from the following:
Wooden chair; brass screw; eraser; cork; iron nail; glass tube; needle; razor-blade; copper wire; iron rod; aluminium saucepan; screwdriver; safety-pin; plastic comb; penknife; shirt; newspaper; handkerchief; bicycle; pencil.

2 Making Magnets

1. Complete these sentences:
(a) Small magnets are made by ———— with another magnet and by the ———— of the Earth's magnetic field. (b) A steel knitting-needle placed near to a bar magnet will become ————. (c) ———— magnets are made electrically. (d) Magnets are demagnetized by hammering and ————. (e) If a magnetized bicycle spoke is broken into pieces, each piece behaves as a separate ————. (f) A horseshoe magnet requires only ———— keeper.
2. Make labelled drawings to explain (a) the domain theory of magnetism, (b) the use of keepers.

3 Using Magnets

1. Complete these sentences:
(a) Lodestones are pieces of black iron oxide called ————. (b) At one time sailors used ———— for finding direction. (c) Magnetite is named after ————, in Asia Minor, where it was once mined. (d) Steel magnets were made by Dr. Gilbert, physician to Queen Elizabeth I, who stroked pieces of sword steel with a ————. (e) A compass needle is affected by a ship's ————. (f) The Earth behaves as a large ————. (g) At present, magnetic north is about 6° west of ———— north. (h) The ———— pole of a magnet should be called the north-seeking pole. (i) The south-seeking pole of a magnet points to the Earth's ———— pole. (j) The angular difference between the geographic and magnetic meridians is called the angle of ————.
2. Write down the names of *six* common pieces of equipment in which magnets are used.

4 Electromagnets

1. Answer these questions (with full sentences):
(a) What happens to a compass needle when it is held close to a wire through which a heavy electric current is flowing? (b) What is the name of the Danish scientist who noticed that a magnet was affected by the electric current flowing in a nearby coil of wire? (c) What determines the polarity of the end of a solenoid when a current is flowing? (d) How can a large iron nail be softened? (e) What is a permanent magnet? (f) What determines the strength of an electromagnet? (g) What is the advantage of using soft-iron cores in electromagnets?
2. Make drawings to explain the polarity rule for a solenoid.

5 Using Electromagnets

1. Write down the names of *five* common pieces of equipment in which electromagnets are used.
2. Make a labelled drawing of an electric bell.

6 Electricity

1. Write a few sentences about each of these topics:
(a) Frictional electricity. (b) Dr. Gilbert. (c) Electroscope. (d) Thunder and lightning. (e) Lightning conductor. (f) Benjamin Franklin. (g) Leyden jar.
2. Write sentences to show the meanings of these words: Static; elektron; amber; conductor; discharge; capacitor.
3. Make a labelled sectional drawing of a Leyden jar.

7 Electric Cells

1. Arrange these jumbled words to make proper sentences:
(a) electricity not stored frictional is easily (b) the simple poles of are copper zinc and a cell plates (c) electrodes the a battery plates of called are (d) hydrogen carries charges the copper positive to plate (e) electrode of cell positive a the Leclanché is rod surrounded dioxide carbon by manganese a (f) layer a develop in polarization not Leclanché does a cell.
2. Write a few sentences to explain what is meant by (a) local action, (b) polarization, (c) potential difference.
3. Make labelled sectional drawings of (a) a Leclanché cell, (b) a dry cell.

Exercises

8 Electric Circuits

1. Complete these sentences:
(a) When a circuit is broken, the __________ ceases to flow. (b) Materials which carry electricity are called __________. (c) Materials which do not carry electricity are called __________. (d) Silver and __________ are good conductors. (e) Dry air and pure water are __________ insulators. (f) Water containing impurities is a __________. (g) Electrical equipment and wires are often covered with __________ materials. (h) The enamel and __________ insulation on the end of a cable must be removed before a connection is made. (i) Telegraph wires are carried on __________ insulators. (j) Electric cables are supported on __________. (k) The __________ itself is used as a conductor in an earth return circuit. (l) Switches have __________ or porcelain covers.
2. Make *two* separate lists of conductors and insulators from the following materials:
Silver; glass; rubber; wood; gold; copper; paper; cotton; aluminium; zinc; plastic; lead; pure water; brass; impure water; porcelain; graphite; oil; tin; dry air; iron; cork; coke; bronze.
3. Some calculations:
(a) What voltage is supplied by a battery of 40 dry cells connected in series? (b) What is the potential difference of 10 dry cells connected in series? (c) 4 dry cells connected in series supply 4 lamps connected in parallel. What is the potential difference across each lamp? (d) 2 dry cells connected in series supply 3 lamps connected in series. What is the potential difference across each lamp? (e) How many dry cells connected in series would be required to supply four 3-volt lamps connected in series?

9 Using Electricity

1. Write brief explanations of these statements:
(a) Insulators have very high resistances. (b) Lamp bulbs are filled with inert gases. (c) Twenty 12-volt lamps connected in series can be operated safely on a 240-volt mains supply. (d) One side of every lamp in a motor car is connected to the frame. (e) Filaments and elements are made of metals with high melting-points.

10 Some Electrical Models

1. Make labelled drawings of (a) model house-lighting circuit, (b) two-way switch circuit.
2. (a) Make a large drawing to show how you would make the connections in a series circuit to illuminate a stage set containing the following: Moon; street lamp; two motor-car headlamps; policeman's lamp; light from a window. What would be the voltage of the battery used with 3-volt lamps? (b) Now make a drawing to show the lamps connected in parallel. What would be the voltage of the battery used?

11 Measuring Heat

1. Write brief explanations of these statements:
(a) The cool water in a swimming bath contains more heat than a teaspoonful of boiling water. (b) The ability of a body to hold heat is called its heat capacity. (c) The heat capacity of a lump of iron is 1/9 that of the same weight of water.
2. Make labelled drawings to show how land and sea breezes are caused by the high heat capacity of water.
3. Some calculations:
(a) Two beakers containing water are heated over the same bunsen flame for the same length of time. The temperature rise of the 500 g of water in one beaker is three times that of the water in the other beaker. What mass of water is contained in the other beaker? (b) How many times greater is the heat capacity of $2\frac{1}{2}$ kg of water than the heat capacity of $\frac{1}{2}$ kg of water? (c) What quantity of heat, in joules, is required to raise the temperature of 90 g of iron from 10°C to 20°C? (d) How much heat, in joules, is required to raise the temperature of (i) 50 g of water through a range of 17°C, (ii) 10 g of water from 25°C to $33\frac{1}{2}$°C? (e) How much heat, in kilojoules, is required to raise the temperature of (i) 10 kg of water through a range of 20°C, (ii) 10 g of water from 0°C to 100°C? (f) In 2 minutes the temperature of 200 g of water is raised from 16°C to 64°C by a gas flame. How much heat, in joules, is supplied by the gas flame in 1 minute? (g) How much heat is required to boil 1000 kg of water if its initial temperature is 40°C?

12 Latent Heat

1. Answer these questions:
(a) What name is given to the heat used in changing water into water vapour or ice into water? (b) What are the three states in which matter exists? (c) What is meant by latent specific heat of fusion? (d) What is meant by specific latent heat of evaporation? (e) What name is given to the tiny particles of which matter is composed? (f) What happens to the molecules of a liquid when it is strongly heated? (g) What is the specific latent heat of steam? (h) What is the specific latent heat of ice? (i) How does perspiring keep a person cool in hot weather? (j) Why is water used in fire-fighting?
2. Write sentences to show the meanings of these words:
Latent; fusion; molecules; vapour; gas; volatile; perspiration; scald.

Exercises

3. Write down the names of a few (a) substances which can be easily vaporized, (b) real gases.

4. People sometimes cool lemonade, beer and other drinks by adding lumps of ice. Write a few sentences to explain why the ice melts and how the drink is cooled.

13 Refrigeration

1. Correctly pair these groups of words together to form proper sentences:

(a) Bacteria cause food	under cold conditions.
(b) Food is preserved by	in refrigerator pipes.
(c) Bacteria cannot reproduce	when it is cooled.
(d) Ammonia is often used	called "dry ice".
(e) Compressed air liquefies	to become tainted.
(f) Liquid air is stored in	canning, bottling and cold storage.
(g) Solid carbon dioxide is	from solid to vapour.
(h) Naphthalene evaporates directly	thick-walled metal cylinders.

2. Write brief explanations of these statements:
(a) On a mountain top, water boils at a temperature lower than 100°C. (b) Naphthalene moth-balls slowly become smaller as they are used. (c) Antifreeze mixtures are put into car radiators during cold spells.

3. Make a labelled sectional drawing to show the main working parts of a refrigerator.

4. Consider the refrigerator you have drawn and then state the purposes of the following:
(a) Tank. (b) Freezing unit. (c) Freezing tube. (d) Pump. (e) Cooling vane. (f) Cooling tube. (g) Valve.

14 The Weather

1. Complete these sentences:
(a) Climate is the ——— state of the weather. (b) California has a warm, ——— climate. (c) A very ——— atmosphere contains much water vapour. (d) The best conditions for rapid ——— are warm surroundings, a dry atmosphere and a large exposed surface. (e) Evaporation is quicker on warm days than on ——— days. (f) Evaporation from the large ——— of a pond is quicker than from a deep tank. (g) The best conditions for condensation are a damp atmosphere and ——— surroundings. (h) ——— easily occurs in an atmosphere containing much water vapour. (i) Clouds are formed when water vapour in the atmosphere is ——— by contact with cold air. (j) On meeting ——— ground, water vapour condenses to form mists. (k) Fogs are dense ——— which have been dirtied by smoke and other impurities. (l) Windows "steam over" in cold weather; water vapour in the warm air of a room ——— on the cold windows., (m) On cold summer mornings, grass is covered in ——— given off by the plants. (n) Dew cannot ——— quickly in cold air. (o) The temperature at which ——— evaporates is called the dew-point. (p) The dew-point varies with the ——— of the atmosphere. (q) Fogs and ——— are a danger to transport. (r) Fires are sometimes used to disperse ———.

15 Winds and Rain

1. Correctly pair these groups of words together to form proper sentences:

(a) Winds are currents	and rises.
(b) Air movements are caused by	high-pressure regions.
(c) Warm air expands	region of high pressure.
(d) Air is heavy in	called cyclones.
(e) Air is light in weight	the approach of a depression.
(f) A depression is a moving	of moving air.
(g) Depressions are sometimes	differences in air pressure.
(h) A fall in the barometer shows	in low pressure regions.
(i) An anticyclone is a moving	region of low pressure.
(j) Anticyclones often bring	heat waves in summer.

2. Write a few sentences to describe the kind of weather usually associated with (a) depressions, (b) anticyclones.

3. Make simple labelled drawings to show how temperature differences on the Earth's surface are caused by (a) the Earth's shape, (b) the atmosphere, (c) the high heat capacity of water.

4. Write sentences to show the meanings of these words:
Anticyclone; meteorologist; muggy; drizzle; windward; regelation; symmetrical; frost; icicle; thaw; hail.

5. Write brief explanations of these statements:
(a) Air moves from high-pressure regions to low-pressure regions. (b) A land surface at the Equator receives more heat from the Sun than a land surface of the same size at the Polar regions. (c) Snowballs are held together by regelation. (d) Thaws have a cooling effect.

6. Make a rough sketch of a snowflake to show the symmetrical arrangement of its crystals.

Exercises

16 Weather Recording

1. Write an essay with the title *Weather Recording*. Expand these paragraph notes. Include a few simple drawings.
Forecasts. Meteorological Office. Weather stations. Forecasts. Farmers, seamen and airmen. Gale warnings.
Conditions. Constantly changing in Britain. Forecasts not completely reliable. Weather sayings often unreliable.
Maps. Daily issue by Met. Office. Isobars. Temperatures as numbers. Directions and strengths of winds. Stations shown by circles. Kind of weather—thunder, sky overcast, hail, etc.
Instruments. Rain-gauge. Thermometer. Hygrometer. Sunshine recorder. Anemometer. Barometer. Weather-vane. Wind-sock.
Recording. Charts, Daily observations of rainfall, temperature, pressure, etc.
Clouds. Cirrus. Cumulus. Stratus. Nimbus. Other cloud types.
2. Write brief explanations of these weather sayings. Are they reliable?
(a) Fine before seven, rain before eleven. (b) The north wind doth blow and we shall have snow. (c) Halo round the moon, bad weather soon. (d) Red sky at night, shepherd's delight. (e) Red sky at morning, shepherd's warning. (f) The east wind is no good for either man or beast.

17 Energy and Work

1. Write a few sentences about each of these topics:
(a) Force. (b) Work. (c) Inertia. (d) Kinetic energy. (e) Potential energy.
2. Write down *five* common examples of (a) kinetic energy, (b) stored energy.
3. Complete these sentences:
(a) Burning coal supplies ———— energy which can be turned into work. (b) Work = effort × ———— through which the effort moves. (c) ———— is measured in joules. (d) More effort is required to move a ———— object than a light object. (e) A person standing in a bus moves forward quickly when the bus stops suddenly; his ———— carries him forward. (f) Moving objects on the Earth come to rest, even when no force is apparently applied to them, because of ———— and friction. (g) More effort is applied by a locomotive in starting a train than in keeping it in ————.
4. Some calculations (assume 1 kgf = 10 N):
(a) How much work, in joules, is done when a load of 10 kgf is lifted for a distance of (i) 3 m, (ii) 6 m, (iii) 50 cm? (b) How much work, in joules, is done when loads of (i) 3 kgf, (ii) 10 kgf, (iii) 0.5 kgf are raised 10 m from the ground? (c) 1000 J of work are required to raise an object 10 m. What is the weight of the object? (d) Through what distance will 1000 J of work raise an object weighing 5 kgf?

18 Movement and Friction

1. Answer these questions:
(a) Why does a bullet, which weighs so little, easily penetrate a thick target? (b) What is the product of the mass and the velocity of a moving object? (c) Why is a piece of string easily broken when its ends are pulled quickly in opposite directions? (d) Why can a tiny hailstone falling from a great height do some damage? (e) Why does an object moving on a rough surface come to rest more quickly than an object moving on a smooth surface? (f) How is friction overcome in machinery? (g) Why are fines thrown on to icy railway lines, do you think?
2. Write a few sentences about each of these topics:
(a) Velocity. (b) Average speed. (c) Momentum. (d) Friction. (e) Lubricants.
3. Write down the names of *three* common devices in which friction is (a) a disadvantage, (b) used to advantage.
4. Some calculations:
(a) What is the speed of a train which travels 90 km in $1\frac{1}{2}$ h? (b) What is the speed of an aeroplane which travels 300 km in 30 min? (c) A man walks 1 km in 15 min. What is his speed in km/h? (d) How far does a cyclist travel in 15 min if his speed is 30 km/h? (e) A marble rolls 10 m in 5 s. What is its average speed in m/s? (f) How long does a marble take to travel 4 m if it is moving at a speed of 2 m/s? (g) How far away is lightning if the thunder is heard 10 s after the flash is seen? Assume that the speed of sound is 300 m/s. (h) A bullet weighing 10 g (10 g = 0.01 kg) has a velocity of 1000 m/s. What is its momentum?

19 Levers

1. Complete these sentences:
(a) The pivot on which a lever turns is called the ————. (b) The turning effect of a lever is called the ————. (c) The work done by a lever = load × distance of the load from the ————. (d) A seesaw is a ————. (e) Mechanical advantage = load ÷ ————. (f) A lever with a large mechanical advantage allows a large ———— to be moved by a small effort. (g) A thin iron bar laid across a beam can be bent fairly easily; the beam gives ————. (h) Devices which contain ———— levers are called double-levers. (i) Nutcrackers and ———— are examples of double-levers. (j) In a rowing boat, the fulcrum is at the place where the oar touches the ————.

Exercises

2. Copy the moments table shown below. Calculate and insert the missing values.

Left-hand side			Right-hand side			Mechanical Advantage
load	distance	moment	effort	distance	moment	load ÷ effort
3 N		300 N cm		50 cm	300 N cm	$\frac{1}{2}$
6 N	40 cm	240 N cm	12 N	20 cm		$\frac{1}{2}$
40 N		2000N cm	10 N	200 cm	2000N cm	
7 N	30 cm		1 N		210 N cm	
20 N				120 cm		2

3. Make drawings of the following common levers. Use arrows and labels to show the positions and the directions of the loads, efforts and fulcra.
(a) Scissors. (b) Wheelbarrow. (c) Nutcrackers. (d) Sugar tongs. (e) Oar of a rowing boat. (f) Pump handle.
4. Make labelled drawings of (a) a Roman steelyard, (b) a Danish steelyard.
5. Some calculations:
(a) In a lever in which the fulcrum is between the load and the effort, a load of 10 kgf is at a distance of 2 m from the fulcrum. What is the distance between the fulcrum and an effort of 5 kgf? (b) A man weighing 70 kgf sits on a seesaw at a distance of 2 m from the pivot. What must be the weight of a boy sitting at a distance of 5 m from the pivot if the seesaw is to balance? (c) A simple lever moves a load of 80 kgf with an effort of 16 kgf. What is the mechanical advantage of the lever? (d) The effort on a lever at a distance of 10 cm from the fulcrum just moves a load at a distance of 50 mm from the fulcrum. What is the mechanical advantage of the lever? (e) The moment of the force on a lever is 40 kgf m units. If the load weighs 8 kgf, what is the distance, in metres, between the load and the fulcrum? (f) An object weighing 10 kgf is attached to the hook of a Roman steelyard. The distance between the hook and the fulcrum is 20 mm. The steelyard is balanced by a sliding weight of 1 kgf. What is the distance between the sliding weight and the fulcrum? (g) The fixed weight on a Danish steelyard which is 40 cm long is 6 kgf. What load on the hook will balance the steelyard when the fulcrum is 10 cm from the fixed weight?

20 Machines

1. Write a few sentences about each of these topics:
(a) The three uses of machines. (b) The mechanical powers. (c) Inclined plane. (d) Efficiency. (e) Wedge. (f) Screw.
2. Write sentences to show the meanings of these words:
Inclined; plane; ramp; axle; vertical; triangular; thread; pitch; slot.
3. Make simple drawings to show how the inclined plane is used in (a) a railway incline, (b) earth ramps of the kind which were probably used in building the Pyramids.
4. Make a labelled drawing of a screw to show what is meant by *thread* and *pitch*.
5. Write down the names of the mechanical powers which are used in each of these devices:
(a) Crowbar. (b) Loading ramp. (c) Bicycle crank. (d) Ship's gangway. (e) Chisel. (f) Axe. (g) Lifting-jack. (h) Railway incline. (i) Door-stop. (j) Scissors. (k) Knife. (l) Hammer. (m) Corkscrew. (n) Staircase. (o) Pump handle. (p) Incisor tooth. (q) Nail. (r) Wheelbarrow. (s) Poker. (t) Ship's propeller.
6. Some calculations:
(a) A 100 kgf sack is raised to the top of a loading ramp 2 m high by an effort of 25 kgf. What is the length of the ramp? (b) What effort, in kgf, is used by a motor car weighing 4000 kgf if it rises 20 m in travelling 2 km? (Neglect the effort required to overcome friction.) (c) The useful work done by a machine is 2000 J and the work put in is 2500 J. What is the efficiency of the machine? (d) How much useful work is done by a machine which is 85% efficient if the work put in is 1000 J?

21 Wheels and Pulleys

1. Write a short essay with the title *The Wheel.* Make your own paragraph notes in the manner shown in previous exercises.
2. Write brief explanations of these statements:
(a) The wheel and axle is a rotating lever. (b) The penny-farthing bicycle is a wheel and axle used for changing slow motion into rapid motion. (c) A single pulley system has no mechanical advantage. (d) In a two-pulley system, one moving and one fixed, there is a mechanical advantage of 2.
3. Make simple labelled drawings of (a) a wheel and axle, (b) a capstan, (c) a single pulley system, (d) a two-pulley system, one moving and one fixed, (e) a driving belt, (f) a reversing belt, (g) a coupling rod and wheels, (h) a gear-wheel.
4. Some calculations:
(a) A wheel has a radius of 10 cm and its axle has a radius of 2 cm. What effort applied to the wheel will raise a load of 50 N attached to the axle? What is the mechanical advantage of the machine? (b) An effort of 4 N applied to a winch arm raises a load of 24 N. What is the length of the winch arm if the load drum has a radius of 2 cm? (c) What effort must be applied to a two-pulley system, one moving and one fixed, to move a load of 28 N? (d) A driving wheel has 84 cogs and its driven wheel has 12 cogs. What is the ratio of the gear-system?

Exercises

22 Force and Strength

1. Complete these sentences:
(a) Forces cause materials to move and to ________ their shapes. (b) The materials used in buildings must be strong enough to withstand large ________. (c) When a material is compressed or pulled apart by two equal, opposing forces, it is said to be under a ________. (d) The change of shape of a material caused by a stress is called a ________. (e) The force required to break a bar is called its ________ stress. (f) When the ends of a material are pushed together by two equal, opposing forces, the stress produced is called ________. (g) When the ends of a material are pulled apart by two equal, opposing forces, the stress produced is called ________. (h) Crane ropes are made of metal wires which withstand ________. (i) The breaking stress of a thread or a wire under tension depends upon its cross-sectional area and the ________ of which it is made. (j) Large buildings are supported by reinforced columns which withstand ________. (k) The bars in metal bridges are arranged into ________ shapes.
2. Write brief explanations of these statements:
(a) Bicycles are made of metal tubes. (b) Buffers are damaged when they are struck by a fast-moving locomotive. (c) Girders are made of strong metals which bend slightly under heavy loads.
3. Make simple drawings to show how the shape of a material is altered by (a) tension, (b) compression.
4. Make simple labelled drawings to show how loads are supported by (a) humpbacked bridges, (b) flat bridges.

23 Power and Engines

1. Answer these questions:
(a) How is power defined? (b) What is the equivalent of 1 watt in joules per second? (c) What was man's first source of power? (d) What are the main sources of power nowadays? (e) What is likely to be the main source of power in the near future? (f) What is the purpose of the flywheel in a steam-engine? (g) Who built the first practicable internal combustion engine? (h) What are the four strokes of a petrol engine? (i) Who invented the diesel engine? (j) Why does a diesel engine require no sparking-plugs?
2. Consider a toy steam-engine and then state the purposes of the following:
(a) Boiler. (b) Safety-valve. (c) Cylinder. (d) Piston. (e) Piston-rod. (f) Crank. (g) Spirit-burner. (h) Flywheel. (i) Exhaust valve. (j) Cylinder pivot.
3. Make simple labelled drawings to show the four-stroke cycle of operations in a petrol engine.
4. If you have an opportunity, make visits to local windmills, water-mills, locomotive-sheds and garages. Make notes and sketches about some of the things you see. Take photographs with your own camera. Obtain permission to do this.
5. Some calculations:
(a) What is the equivalent of 3 kilowatts in (i) N m/min, (ii) N m/s? (b) The power developed by an engine is 182 kgf m/s. What is this in watts? (c) A cyclist, whose speed is constant, travels 1000 m in 2 min. What is the power developed if he applies an effort of 60 N to the pedals?

24 Flight

1. Write sentences to show the meanings of these words: Resistance; parachute; hover; glide; airship; airscrew; compressed; rarefied; lift.
2. Make labelled sectional drawings to show how (a) an aeroplane rises, (b) a kite rises.
3. Write an essay with the title *Flight*. Make your own paragraph notes in the manner shown in previous exercises.
4. Write a few sentences to explain this:
Two apples are suspended from strings so that the distance between the apples is about 1 cm. A boy blows between the apples but he cannot blow them apart.

25 Flying Machines

1. Write a few sentences about each of these topics:
(a) Wilbur and Orville Wright. (b) Helicopter. (c) Firework rocket. (d) Sir Frank Whittle.
2. Write sentences to show the meanings of these words: Plane; fuselage; fabric; bank; rotor; jet.
3. Make a simple labelled sectional drawing to show the main working parts of a jet engine.
4. Consider the controls of an aeroplane and then state the purposes of the following:
(a) Airscrew. (b) Wing. (c) Aileron. (d) Landing flap. (e) Elevator. (f) Rudder.
5. If you have an opportunity, visit an airport or an aerodrome. Make notes and sketches about some of the things you see. Take photographs with your own camera. Obtain permission to do this.

26 The Earth's Crust

1. Complete these sentences:
(a) The inside of the Earth is probably made of iron and ________. (b) The thin outer layer of the Earth is called the ________. (c) Volcanoes erupt hot gases and molten rocks called ________. (d) The ________ was once a ball of burning gases and vapours pulled out of the Sun by the gravitational force of a passing star.

Exercises

(e) —————— ranges and —————— deeps were formed by contractions in the Earth's crust. (f) —————— occur where the Earth's crust is unsteady. (g) —————— are able to decide where coal, oil and metallic ores are likely to be found. (h) Land surfaces are gradually being worn away by —————— agents. (i) Rock surfaces are eaten away by weak —————— in ground water and the atmosphere. (j) The four main weathering agents are rain, ——————, —————— and chemical action. (k) Hot water —————— occur in Iceland, Italy, the United States and New Zealand. (l) Extreme pressure changes soft clay and mud into ——————.

2. State whether these rocks are igneous, sedimentary or metamorphic:
Basalt; sandstone; marble; clay; limestone; coal; granite; quartz; chalk; pumice; slate.

3. Make labelled sectional drawings to show (a) the main parts of the Earth, (b) the three kinds of rocks in the Earth's crust, (c) a volcano.

27 Some Useful Rocks

1. Write a few sentences about each of these topics:
(a) Foraminifera. (b) Marble. (c) Coral. (d) Quicklime. (e) Slaked lime. (f) Mortar. (g) Concrete. (h) Glass.

2. Write brief explanations of these statements:
(a) Marble, limestone, chalk, Iceland spar, calcite and coral are different forms of calcium carbonate. (b) The bricks used in damp courses are made of non-porous clay or are glazed. (c) Mortar becomes very hard over the years.

3. Make a list of *ten* common devices, such as bottles, mirrors, windows, etc., in which glass is used.

28 Metals

1. Complete these sentences:
(a) Ores contain —————— combined with other materials. (b) A few metals, such as copper and ——————, often occur in the pure state. (c) Haematite is an oxide of ——————. (d) Iron oxide + carbon = —————— + carbon dioxide. (e) Metals are good conductors of —————— and electricity. (f) Ornaments are cast in brass and ——————. (g) —————— is a liquid metal. (h) —————— is a soft metal. (i) Iron is not so malleable as ——————. (j) Cold iron is less malleable than —————— iron. (k) An —————— is a solid solution. (l) Brass contains copper and ——————. (m) Bronze contains —————— and tin. (n) Soft solder is a mixture of —————— and lead. (o) —————— is added to silver and gold coins and jewellery to make them durable. (p) —————— is made by adding carbon to iron. (q) Very hard —————— contains as much as 1.5% carbon.

2. Write down the names of *ten* common metals.

3. Write sentences to show the meanings of these words:
Metal; extraction; cast; malleable; ductile; alloy; durable; brittle; tempering; quenching.

4. Write a few sentences to explain the terms *annealing* and *case-hardening*. Use an encyclopaedia.

5. Write down the properties and main uses of each of these metals. Use an encyclopaedia.
(a) Aluminium. (b) Lead. (c) Zinc. (d) Tin. (e) Mercury. (f) Silver. (g) Gold. (h) Copper. (i) Iron.

6. Make a block diagram to show the proportions of iron and carbon in very hard steel. Use a 10 cm square of graph paper with 1 cm squares; this contains 100 small squares. Shade in $1\frac{1}{2}$ small squares to represent carbon.

29 History in the Rocks

1. Write sentences to show the meanings of these words:
Legend; archaeologist; fathom; uranium; fossil; amber.

2. Write an essay with the title *History In The Rocks*. Make your own paragraph notes in the manner shown in previous exercises. Include a few drawings in the essay.

3. Use tracing paper to make copies of some of the extinct animals shown in this chapter.

30 Evolution

1. Write a few sentences about each of these topics:
(a) Evolution. (b) Adaptation. (c) Natural selection. (d) Inheritance. (e) Selective breeding. (f) Reversion to type.

2. Write *two* sentences about each of these animals:
(a) Penguin. (b) Bulldog. (c) Lung-fish. (d) Horse. (e) Dairy cow. (f) Duck-billed platypus. (g) Frog. (h) Lemur. (i) Gorilla. (j) Man.

3. Use tracing paper to make copies of some of the animals shown in this chapter.

31 Soil

1. Correctly pair these groups of words together to form proper sentences:

(a) Soil contains decaying plant	by weathering agents.
(b) Soil is generally darker	water, frost, heat and wind.
(c) Soil is broken up	gravel, lime, humus and mineral salts.
(d) Weathering agents are	are food for plants.
(e) Spaces in the soil are	cover for young seedlings.
(f) Soil contains sand, clay,	material called humus.

Exercises

(g) Mineral salts in soil water	in colour than subsoil.
(h) Soil provides darkness and	filled with air and water.
(i) Soil protects seedlings	against temperature extremes.

2. Write a few sentences about each of these topics:
(a) Humus. (b) Clayey soil. (c) Sandy soil. (d) Chalky soil. (e) Loam. (f) Flocculence. (g) Earthworms.
3. Make a labelled drawing to show the soil, subsoil and underlying rock layers in the side of a railway cutting.

32 Food for Plants

1. Write sentences to show the meanings of these words:
Residue; peat; fertile; rotation; manure; fertilizer; compost; effervescence; neutralize.
2. Draw a chart to show a four-year crop-rotation plan for roots, wheat, oats and clover.
3. Write down the names of some manures and fertilizers which are (a) vegetable, (b) animal, (c) chemical.
4. Write brief explanations of these statements:
(a) Farmers keep their soil fertile by regular ploughing and crop rotation. (b) Farmers sometimes resort to green manuring. (c) Gardeners "sweeten" acid soils by adding a little lime.

33 The Work of Roots

1. Write a few sentences about each of these topics:
(a) Root cap. (b) Root hairs. (c) Semi-permeable membrane. (d) Osmosis. (e) Terracing. (f) Strip-cropping.
2. Write sentences to show the meanings of these words:
Aerial; hydrotropism; transplanting; seedling; brine; wilts; erosion.
3. Write a short essay with the title *The Work of Roots*. Make your own paragraph notes in the manner shown in previous exercises.

34 The Work of Stems and Leaves

1. Complete these sentences:
(a) Stems hold plants erect so that their leaves can obtain ————. (b) Climbing plants support themselves with ———— and hooks. (c) Plants ———— when they cannot obtain enough water. (d) Stems of herbaceous plants are kept erect by their sap-filled ————. (e) Leaves ————, transpire water, and make and store food. (f) Onion ———— consist mainly of swollen food leaves. (g) In the presence of sunlight, leaves make starch out of carbon dioxide in the atmosphere and water by a process called ————. (h) Photosynthesis only occurs in the presence of a green pigment called ————. (i) Grass growing under a stone is ———— and weakly. (j) Leaves give off ———— during photosynthesis. (k) Plants take in oxygen when they ————. (l) Water plants in an aquarium give off ———— which fish breathe.

35 In the Garden

1. Write sentences to show the meanings of these words:
Pollinate; aerate; segment; wormery; bristle; burrow; pest; fungicide; insecticide.
2. Write a few sentences to explain how each of these animals either helps or hinders the gardener:
(a) Bee. (b) Slug. (c) Ladybird. (d) Earwig. (e) Caterpillar. (f) Spider. (g) Toad. (h) Greenfly. (i) Earthworm. (j) Mouse.
3. Write brief explanations of these statements:
(a) One should not be too ready to regard certain animals as pests. (b) Weasels and stoats were introduced into Australia as a means of destroying rabbits; this attempt at biological control was not successful.

36 Some Tiny Plants

1. Write a few sentences about each of these topics:
(a) Bacteria. (b) Root nodules. (c) Yeast. (d) Bread-making. (e) Fermentation. (f) Alcohol. (g) Methyl alcohol. (h) Vinegar. (i) Beer.
2. Write sentences to show the meanings of these words:
Leguminous; sour; sterilize; bloom; bud; intoxication; distilled; stimulant; solvent; malt.

Answers to Numerical Exercises

Chapter 8: 3. (a) 60 V (b) 15 V (c) 6 V (d) 1 V (e) 8.
Chapter 11: 3. (a) 1.5 kg (b) 5 (c) 400 J (d) (i) 3400 J (ii) 340 J (e) (i) 800 kJ (ii) 4 kJ (f) 19 200 J (g) 240 MJ.
Chapter 17: 4. (a) (i) 300 J (ii) 600 J (iii) 50 J (b) (i) 300 J (ii) 1000 J (iii) 50 J (c) 10 kgf (d) 20 m.
Chapter 18: 4. (a) 60 km/h (b) 600 km/h (c) 4 km/h (d) 7.5 km (e) 2 m/s (f) 2 s (g) 3 km (h) 10 Ns.
Chapter 19: 2.

3 N	100 cm	300 N cm	6 N	50 cm	300 N cm	$\frac{1}{2}$
6 N	40 cm	240 N cm	12 N	20 cm	240 N cm	$\frac{1}{2}$
40 N	50 cm	2000 N cm	10 N	200 cm	2000 N cm	4
7 N	30 cm	210 N cm	1 N	210 cm	210 N cm	7
20 N	60 cm	1200 N cm	10 N	120 cm	1200 N cm	2

5. (a) 4 m (b) 28 kgf (c) 5 (d) 2 (e) 5 m (f) 20 cm (g) 2 kg.
Chapter 20: 6. (a) 8 m (b) 40 kgf (c) 80% (d) 850 J.
Chapter 21: 4. (a) 10 N (b) 12 cm (c) 14 N (d) 7.
Chapter 23: 5. (a) (i) 180 000 N m/min (ii) 3000 N m/s (b) 1820 W (c) 500 W.

Index